Studies in Computational Intelligence

Volume 702

Series editor

Janusz Kacprzyk, Polish Academy of Sciences, Warsaw, Poland
e-mail: kacprzyk@ibspan.waw.pl

About this Series

The series "Studies in Computational Intelligence" (SCI) publishes new developments and advances in the various areas of computational intelligence—quickly and with a high quality. The intent is to cover the theory, applications, and design methods of computational intelligence, as embedded in the fields of engineering, computer science, physics and life sciences, as well as the methodologies behind them. The series contains monographs, lecture notes and edited volumes in computational intelligence spanning the areas of neural networks, connectionist systems, genetic algorithms, evolutionary computation, artificial intelligence, cellular automata, self-organizing systems, soft computing, fuzzy systems, and hybrid intelligent systems. Of particular value to both the contributors and the readership are the short publication timeframe and the worldwide distribution, which enable both wide and rapid dissemination of research output.

More information about this series at http://www.springer.com/series/7092

Piotr Hońko

Granular-Relational Data Mining

How to Mine Relational Data in the Paradigm of Granular Computing?

Springer

Piotr Hońko
Faculty of Computer Science
Bialystok University of Technology
Bialystok
Poland

ISSN 1860-949X   ISSN 1860-9503   (electronic)
Studies in Computational Intelligence
ISBN 978-3-319-84977-5   ISBN 978-3-319-52751-2   (eBook)
DOI 10.1007/978-3-319-52751-2

Printed on acid-free paper

This Springer imprint is published by Springer Nature
The registered company is Springer International Publishing AG
The registered company address is: Gewerbestrasse 11, 6330 Cham, Switzerland

If I have seen further, it is by standing upon the shoulders of giants.

Isaac Newton (1642–1727)
Letter to Robert Hooke, 1675

Preface

Relational data mining is application of data mining techniques to discover knowledge that is hidden in data with a relational structure. It aims to integrate methods from existing fields applied to an analysis of data represented by multiple relations, producing new techniques for mining relational data. It has been successfully applied in areas such as bioinformatics, marketing, or fraud detection.

Granular computing is a new and rapidly growing paradigm of information processing. It integrates theories, methodologies, techniques, and tools that make use of granules in the process of problem solving. Granular computing methods have widely and successfully been applied in the field of data mining. They have mainly been used to discover knowledge from single table databases; however, research on incorporating them into mining relational data has also been done.

The goal of this monograph is to highlight research on mining relational data in the paradigm of granular computing. This newly emerging field can be identified as granular computing-based relational data mining and shortly called *granular-relational data mining*. The monograph provides unified frameworks for performing typical data mining tasks such as classification, clustering, and association discovery. The book is also aimed to establish itself as a basic text at the intersection of two fields: relational data mining and granular computing.

The project was partially funded by the National Science Center awarded on the basis of the decision number DEC-2012/07/B/ST6/01504.

Białystok, Poland   Piotr Hońko
September 2016

Contents

Contents

Symbols

A	Set of attributes
A_{des}	Subset of the attribute set consisting of descriptive attributes
A_{key}	Subset of the attribute set consisting of key attributes
$acc_{\mathbf{IS}}(\alpha \leftarrow \beta)$	The accuracy of $\alpha \leftarrow \beta$ with respect to an information system $\mathbf{IS}$
$acc^{\pi_a}_{IS_{(m)}}(\varepsilon_{\alpha \to \beta})$	The accuracy of $\varepsilon_{\alpha \to \beta}$ with respect to a domain of $D_a(\varepsilon_\alpha)$
$AS_{\#,\$}$	Approximation space
$AS^i_{\#,\$}$	Approximation space constructed based on i-th depth level related sets
$AS^{gen}_{i,\#,\$}$	Approximation space constructed based on i-th depth level generalized related sets
AS_ω	(Non-compound) approximation space
$AS_{\omega(i,j)}$	Compound approximation space constructed based on $IS_{(i,j)}$
$AS_{\omega(m)}$	Compound approximation space constructed based on $IS_{(m)}$
$AS^\Theta_{\omega(i,j)}$	Constrained compound approximation space constructed based on $IS^\Theta_{(i,j)}$
$AS^\Theta_{\omega(m)}$	Constrained compound approximation space constructed based on $IS^\Theta_{(m)}$
$conf_{\mathbf{IS}}(\alpha \to \beta)$	The confidence of $\alpha \to \beta$ with respect to an information system $\mathbf{IS}$
$conf^{\pi_a}_{IS_{(m)}}(\varepsilon_{\alpha \to \beta})$	The confidence of $\varepsilon_{\alpha \to \beta}$ with respect to a domain of $D_a(\varepsilon_\alpha)$
$conf^{\pi_a}_{IS_{(m)}}(\varepsilon_{\alpha \to \beta}, t^{a',\#})$	The confidence of $\varepsilon_{\alpha \to \beta}$ with respect to a domain $D_a(\varepsilon_\alpha)$ under a threshold $t \in [0,1]$ imposed on a domain $D_{a'}(\varepsilon_\alpha)$
$cov^{\pi_a}_{IS_{(m)}}(\varepsilon_{\alpha \to \beta})$	The coverage of $\varepsilon_{\alpha \to \beta}$ with respect to a domain of $D_a(\varepsilon_\alpha)$
$D(\varepsilon_\alpha)$	The domain of an ε-relation ε_α
$D_a(\varepsilon_\alpha)$	The domain related to an attribute $a \in attr(\varepsilon_\alpha)$
ε_α	Formula-based relation, i.e., relation defined by a formula $\alpha \in L_{\mathbf{IS}}$
ε_{IS}	Relation representing an information system $IS = (U,A)$

ε−relation	Relation with the schema $\varepsilon_\alpha(a_1, \ldots, a_k)$ constructed based on a formula $\alpha \in L_{IS_{(m)}}$
$freq_{\mathbf{IS}}(\alpha)$	The frequency of α with respect to an information system $\mathbf{IS}$
$freq_{IS_{(m)}}^{\pi_a}(\varepsilon_\alpha)$	The frequency of ε_α with respect to a domain $D_a(\varepsilon_\alpha)$
$freq_{IS_{(m)}}^{\pi_a}(\varepsilon_\alpha, t^{a',\#})$	The frequency of ε_α with respect to a domain $D_a(\varepsilon_\alpha)$ under a threshold $t \in [0, 1]$ imposed on a domain $D_{a'}(\varepsilon_\alpha)$
$I_\#$	Uncertainty function of $AS_\#$
I_ω	Uncertainty function of AS_ω
$I_{\omega(i,j)}$	Uncertainty function of $AS_{\omega_{(i,j)}}$
$I_{\omega(m)}$	Uncertainty function of $AS_{\omega_{(m)}}$
$I_{\omega_{(i,j)}}^{\Theta}$	Uncertainty function of $AS_{\omega_{(i,j)}}^{\Theta}$
$I_{\omega_{(m)}}^{\Theta}$	Uncertainty function of $AS_{\omega_{(m)}}^{\Theta}$
IS	(Non-compound) information system
$IS_{(i,j)}$	Compound information system consisting of particular systems IS_i and IS_j
$IS_{(m)}$	Compound information system consisting of particular systems $IS_1, \ldots, IS_m$
$IS_{(i,j)}^{\Theta}$	Constrained compound information system, i.e., compound information system with the universe constrained by formulas from Θ
$IS_{(m)}^{\Theta}$	Constrained compound information system, i.e., compound information system with the universe constrained by formulas from Θ
L_{IS}	Language for constructing granules in $\mathbf{IS} \in \{IS, IS_{(i,j)}, IS_{(m)}, IS_{(i,j)}^{\Theta}, IS_{(m)}^{\Theta}\}$
$L_{IS_{des}}$	Language for constructing granules based on attributes from A_{des} of IS
$L_{IS_{key}}$	Language for constructing granules based on attributes from A_{key} of IS
$L_{IS_{(i \vee j)}}$	Language consisting of formulas from L_{IS_i} and L_{IS_j}
$L_{IS_{(i \vee j)}^{\Theta}}$	Language consisting of formulas from $L_{IS_i^{\Theta}}$ and $L_{IS_j^{\Theta}}$
$L_{IS_{(i \wedge j)}}$	Language consisting of formulas constructed over both L_{IS_i} and L_{IS_j}
$L_{IS_{(i \wedge j)}^{\Theta}}$	Language consisting of formulas constructed over both $L_{IS_i^{\Theta}}$ and $L_{IS_j^{\Theta}}$
$L_{\mathbf{IS}}^{\star}$	Language $L_{\mathbf{IS}}$ expanded by formulas constructed using ε-relation
$\nu_\$$	Rough inclusion of $AS_\$$
ν_ω	Rough inclusion of AS_ω
$\nu_{\omega_{(i,j)}}$	Rough inclusion of $AS_{\omega_{(i,j)}}$
$\nu_{\omega_{(m)}}$	Rough inclusion of $AS_{\omega_{(m)}}$

$\nu^{\Theta}_{\omega_{(i,j)}}$ Rough inclusion of $AS^{\Theta}_{\omega_{(i,j)}}$

$\nu^{\Theta}_{\omega_{(m)}}$ Rough inclusion of $AS^{\Theta}_{\omega_{(m)}}$

$rlt(o)$ Related set of an object o

$rlt^n(o)$ n-th depth level related set of an object o

$rlt_{gen}(o)$ Generalized related set of an object o

$rlt^n_{gen}(o)$ n-th depth level related set of an object o

$SEM_{\mathbf{IS}}(\alpha)$ The semantics of a formula α in an information system **IS**

$SEM^{\pi_i}_{IS_{(m)}}(\alpha)$ The semantic of α in $IS_{(m)}$ limited to objects from IS_i

$SEM^{\pi_{(i_1,i_2,...,i_k)}}_{IS_{(m)}}(\alpha)$ The semantic of α in $IS_{(m)}$ limited to objects from $IS_{(i_1,i_2,...,i_k)}$

U Universe, i.e., the set of objects

U^i Universe constructed based on i-th depth level (generalized) related sets

U_D Universe constructed based on a relational database D

U_{D_B} Universe constructed based on a relational database D limited to background objects

U_{D_T} Universe constructed based on a relational database D limited to target objects

U^{α} Subset of U including objects satisfying a formula α

$U_{\omega_{(i,j)}}$ Compound universe, i.e., the Cartesian product of universes U_i and U_j

$U_{\omega_{(m)}}$ Compound universe, i.e., the Cartesian product of universes $U_1, \ldots, U_m$

$U^{\Theta}_{\omega_{(i,j)}}$ Constrained compound universe, i.e., the subset of the compound universe consisting of pairs satisfying formulas from Θ

$U^{\Theta}_{\omega_{(m)}}$ Constrained compound universe, i.e., the subset of the compound universe consisting of m-arity tuples satisfying formulas from Θ

Chapter 1
Introduction

One of the main challenges of data mining is to develop its unifying theory [102]. The current state of the art of data mining research seems too ad hoc. Many techniques are designed for individual problems, such as classification, clustering, or association discovery, but there is no unifying theory. A theoretical framework that unifies different data mining tasks can help the field and provide a basis for future research.

The problem of developing a unified framework is more complicated for relational data than for that stored in a single table. Such data is distributed over multiple tables, and the central issue in the specification of a relational data mining problem is the definition of a model of the data. Such a model directly determines the type of patterns that will be considered, and thus the direction of the search. Such specifications are usually referred to as declarative or language bias [49]. A bias not only determines pattern structure, but also limits the search space which can be very huge for relational data.

The unification in the field of relational data mining can be done by building a bridge between relational data and knowledge to be discovered from it. Granular computing as a paradigm of information processing has shown to be a proper environment for building such a construction. A granular representation of relational data can be seen, on one hand, as an alternative view of considered objects, and on the other hand, as a platform for discovering relational patterns of different types.

This monograph aims to provide comprehensive frameworks to mining relational data in the paradigm of granular computing. These two fields, i.e. relational data mining and granular computing are more particularly described in the current chapter. Each of the two parts of the monograph concerns one general granular computing approach for mining relational data. Part I describes a generalized related set based approach and is structured as follows. Chapter 2 introduces an information system dedicated to relational data. Information granules in this system are defined based on the notion of generalized related sets that are the basis to discover knowledge from relational data. Chapter 3 investigates properties of the granular computing frame-

© Springer International Publishing AG 2017
P. Hońko, *Granular-Relational Data Mining*, Studies in Computational
Intelligence 702, DOI 10.1007/978-3-319-52751-2_1

work. The properties enable to improve the performance of tasks such as relational objects representation, search space limitation, and relational patterns generation. Chapter 4 provides specialized versions of the granular computing framework for association discovery and classification rule mining. Chapter 5 develops a relational version of rough-granular computing by defining approximation spaces and rough approximations for relational data.

Part II presents a description language based approach and is structured as follows. Chapter 6 introduces compound information systems and their constrained versions. They are an extension of the standard information system to a relational case. Chapter 7 shows how the granular computing based framework defined for single table data can be upgraded to a relational case. Chapter 8 develops relation-based granules that make it possible to discover richer knowledge from relational data. Chapter 9 provides compound approximation spaces that are intended to handle uncertainty in relational data.

Conclusions on application of granular computing to relational data mining are given in Chap. 10.

1.1 Relational Data Mining

Multi-relation data mining (MRDM) (see, e.g. [12, 18, 25, 30, 53]) concerns knowledge discovery from relational databases consisting of multiple relations (tables). Research on relational data covers a range of data mining tasks such as classification (see, e.g. [13, 77, 94, 108]), clustering (see, e.g. [11, 14, 29, 47]), association discovery (see, e.g. [7, 20–22]), subgroup discovery (see, e.g. [55, 56, 99, 111]), sequence mining (see, e.g. [17, 27, 28, 46]), outlier detection (see, e.g. [3, 64, 78, 79]). Mining relational data has been supported by standard data mining techniques twofold: upgrading propositional algorithms to a relational case (see, e.g. [13, 70, 77, 95]) and using propositional algorithms to relational data transformed beforehand into a single table (see, e.g. [4, 50, 51, 54]).

Relational data is usually mined using one of the general frameworks: inductive logic programming and relational database theory.

Early approaches for pattern discovery in relational data were defined in an inductive logic programming (ILP) framework [24, 26, 69]. ILP is a research field at the intersection of machine learning and logic programming. It provides a formal framework as well as practical algorithms for learning in an inductive way relational descriptions from data represented by target examples and background knowledge.

In ILP, data and induced patterns are represented as formulas in a first-order language. Data is stored in deductive databases, where relations can be defined extensionally as sets of ground facts and intentionally as sets of database clauses. Patterns are typically expressed as logic programs, i.e. sets of Horn clauses.

In ILP, the pattern structure is determined by the so-called declarative bias. It imposes some constraints on the patterns to be discovered. Thanks to the bias, one can determine which relations and how many times may be used in patterns; how to replace a relation attribute with a variable; what values a relation variable may take, and the like.

An alternative framework [48, 49] for discovering patterns in relational data is defined in relational database theory (RDB). In relational database, relations are usually defined extensionally as sets of tuples of constants. However, they can also be defined intensionally as sets of views. Relational patterns discovered in relational database can be expressed as SQL queries.

Unlike in the ILP framework, a specification of the pattern structure is not required. Instead, the patterns are specified by the relationships that occur between the database entities and are shown by entity-relationship diagram. Alternatively, class diagram that is a part of Unified Modeling Language (UML) [49] is used to express a bias. UML class diagram shows how associations (i.e. structural relationships) between given classes (which correspond to database tables) determine how objects in each class relate to objects in another class. Furthermore, multiplicities of associations are also considered. Such an association multiplicity provides information how many objects in one class are related to a single object in another, and vice versa.

1.2 Granular Computing

When analyzing data to discover knowledge, regardless of the tool used, we usually aggregate the objects with common features into the same clusters (i.e. groups). Such clusters can be treated as information derived from the database, which is, in turn, the basis for the discovery of knowledge. The clusters can be obtained in a variety of ways depending, among others, on the task to be performed. Furthermore, one can receive many different partitions of the universe, i.e. families of clusters, for the same task. The choice of the most proper partition can depend on which solution accuracy of the problem under consideration is sufficient. The challenge is thus to develop a framework for constructing and processing such clusters of data.

A field within which frameworks are developed for problem solving by the use of granules (e.g. clusters of data) is granular computing (GC) [9, 72]. This is a relatively new, rapidly growing field of research (see, e.g. [5, 6, 16, 23, 45, 57, 61, 63, 73, 86, 97, 100, 104]). It can be viewed as a label of theories, methodologies, techniques, and tools that make use of granules in the process of problem solving [105].

A granule is a collection of entities drawn together by indistinguishability, similarity, proximity or functionality [110]. Therefore, a granule can be defined as any object, subset, class, or cluster of a given universe. The process of the formation of granules is called granulation. To clearly differentiate granulation from clustering, the semantic aspect of GC is taken into account. Therefore, we treat information granulation as a semantically meaningful grouping of elements based on a given criterion [10]. An information granule can be represented by an expression of the form $(name, content)$, where $name$ is the granule identifier and $content$ is a set of objects identified by $name$ [89].

Granulation can be performed by applying a top-down or a bottom-up method. The former concerns the process of dividing a larger granule into smaller and lower level granules, and the latter the process of forming a larger and higher level granule with smaller and lower level sub-granules [103].

One can obtain many granularities of the same universe which differ in their levels. A granule of high-level granularity, i.e. a high-level granule represents a more abstract concept, and a low-level granule a more specific one. A basic task of GC is to switch between different levels of granularity. A more specific level granularity may reveal more detailed information. On the other hand, a more abstract level granularity may improve a problem solution thanks to omitting irrelevant details.

1.3 Granular Computing Tools: Rough Set Theory

Handling uncertainty in data is a challenging task in the field of data mining. A powerful framework intended for this issue is provided by rough set theory [71]. It was proposed by Professor Zdzisław Pawlak in early 1980s as a mathematical tool to deal with uncertainty in data. Although being a standalone field, rough set theory, beside fuzzy set theory [109], is considered as one of the main granular computing tools. A concept that can include uncertain data is characterized in this theory by a pair of two certain sets, i.e. its lower and upper approximations. New knowledge about the concept can be derived from the approximations. For example, decision rules constructed based on the lower approximation show features of objects that certainly belong to the concept, whereas those generated from the upper one describe objects whose membership in the concept is possible.

Many various rough set models have been proposed over the last three decades (e.g. [32, 82, 107, 116]). They have found a wide range of applications in areas such as e.g. medicine, banking, or engineering (for more details, see, e.g. [74, 81, 87]). The standard rough set model has been generalized in a variety of ways. However, most of the rough set approaches are intended to analyze data stored in a single table. Such a data structure makes it possible to encode simple features of objects of the interest. To show more complex properties such as relationships among objects, a more advanced structure is needed, e.g. relational database.

An adaptation of rough sets tools to relational data can be considered as a generalization of the standard rough set model. An overview of this direction of development of the field is provided below.

The standard rough set model was extended to a covering generalized rough set model (e.g. [15, 115]), where the universe is replaced with its covering. Such a generalization enables to deal with more complex problems. Covering rough set theory with the concept of neighborhood induced by covering plays an important role in reduction of nominal data and in generation of decision rules from incomplete data.

In [62] a relationship between different approximation operators defined in covering rough set theory was studied. It was shown that the operators that use the notion of neighborhood and the complementary neighborhood can be defined almost in the same way. It was also investigated that such twin approximation operators have similar properties.

In [91] matrix-based methods for computing approximations of a given concept in a covering decision system is proposed. The methods are also used for reducing covering decision systems. It was shown that the proposed approach can decrease the computational complexity for finding all reducts.

Another generalization of the standard rough set model (single granulation rough set model) is rough set model based on multi-granulations (MGRS) [76]. Approximations of a concept are defined by using multiple equivalence relations on the universe. The relations are chosen according to user requirements or the problem to be solved. MGRS is considered in two different versions. If the condition of the lower approximation is satisfied for (at least one of/all) single granulation rough set models under consideration, then MGRS is called (optimistic/pessimistic). Properties of optimistic and pessimistic multi-granulation rough set models investigated in [80] show connections of these models with notions such as lattices, topology on the universe, and Boolean algebra.

A model that can be viewed as a multi-granulations form of nearness approximation space was introduced in [96]. A topological neighborhood based on *EI* algebra (a notion from axiomatic fuzzy set theory) is used in information systems with many category features. The neighborhood is combined with generalized approximation spaces producing, thereby, an extension model of the approximation space.

Another kind of extension of the standard rough set model is composite rough set model [112] that is intended for dealing with multiple types of data in information systems, e.g. categorical data, numerical data, set-valued data, interval-valued data and missing data. All basic rough set notions such as lower and upper approximations, positive, boundary and negative regions are redefined in composite information systems. A dynamic version of the composite rough set model [113], which uses a matrix-based method, makes it possible to fast update approximations of a changing concept.

1.4   Mining Relational Data Using Granular Computing

The development of general granular computing frameworks for mining relational data is a relatively new direction. Instead of approaches described in this book one can point out the following ones.

An information system dedicated to relational data, which is defined in a granular computing environment, is proposed in [84]. The information system, called a sum of information systems, is the pair of the universe (the Cartesian product of the universes of the information systems, each corresponding to one table) and the attribute set (the collection of attributes from the attribute sets of the information systems) A constrained version of this system allows only tuples of objects that belong to a constraint relation on the Cartesian product of the universes. The constraint relation can be constructed by conditions expressed by Boolean combination of descriptors of attributes. Not only the attributes from the attribute set, but also some other ones specifying relation between particular information systems can be used to define the constraints. Such systems can find application in reasoning in distributed systems of granules and in searching for patterns in data mining.

Another approach where relational data is viewed in the context of GC is introduced in [58]. A relational database can be represented by a relational granular model which is the pair of the universe (a family of classical sets, called the family of universes) and the collection of relations on the Cartesian product of sets from the universe. The sets from the universe correspond to objects of a relational database, and the relations (of various arities) define constraints for these objects. Some granules considered in fields such as data mining, web/text mining, and social networks can be modeled into the relational granular model.

More research has been done for applying one of the main tools of GC, i.e. rough set theory to relational data.

In [101] the approximation space is defined as a triple of two distinct universes and a binary relation that is a subset of the Cartesian product of the universes. Approximations are defined for a subset of one of the universes. They include objects from the other universe that are in the relation with objects of the subset. Such an approach can be viewed as a generalization of that introduced in [106] where approximations are defined in a formal context that is a triple of a universe of objects, universe of attributes, and a binary relation between the universes.

In [52] approximations are defined in an information system that is a pair of the double universe (the Cartesian product of two particular universes) and the attribute set. Approximations of a subset of the double universe are defined based on equivalence classes of the equivalence relation on the double universe. Additionally, a constrained version of the information system is introduced. It is a triple of the double universe, a constraint relation on the universe, and the attribute set.

To handle with data stored in many tables a multi-table information system is proposed in [68]. The system is a finite set of tables (each table is viewed as an information system). Approximations are defined for a subset of the universe of one specified table, i.e. the decision table. Elementary sets of a given universe are used to define the approximations. Indiscernibility of objects from the decision table is defined using the information available in all the tables of the multi-table information system.

The general granular computing approach from [84] was also used to process relational data using a rough set approach. Approximation spaces for multiple universes are constructed based on (constraint) sums of information systems. Approximations are defined for a subset of the Cartesian product of the universes using approximations computed for particular information systems.

Part I
Generalized Related Set Based Approach

Chapter 2
Information System for Relational Data

2.1 Introduction

The goal of this chapter is to develop a general granular computing based framework for mining relational data. It is based on an information system defined for relational data [38]. Information granules derived from the information system are defined based on the notion of related sets, that is sets of objects related (i.e. joined) to the objects to be analyzed. Such granules are the basis for discovering relational knowledge.

The crucial task of the general framework is to process relational data for discovering patterns of different types. Namely, information granules obtained in the framework can be viewed as an abstract representation of relational data. Such a representation is treated as the search space for discovering relational patterns. Thanks to this, the size of the search space may be significantly limited.

The framework is independent on the way the language bias is specified, thereby biases from existing frameworks can be adapted. Furthermore, the framework, unlike others (i.e. ILP, RDB), unifies not only the way the data and patterns are expressed and specified, but also partially the process of discovering patterns from the data. Namely, the patterns can directly be obtained from the information granules or constructed based on them.

Applying the granular computing idea makes it possible to switch between different levels of granularity of the same universe (i.e. the set of objects), thereby one can choose an appropriate granularity of the data for a given task.

In the framework, one can define new methods as well as redefined existing ones for performing popular relational data mining tasks.

P. Hońko, *Granular-Relational Data Mining*, Studies in Computational Intelligence 702, DOI 10.1007/978-3-319-52751-2_2

The remaining of the chapter is organized as follows. Section 2.2 constructs an information system for relational data. Section 2.3 defines a granular description of relational objects that is based on the notion of generalized related sets. Section 2.4 shows how to construct relational patterns based on introduced granules. Section 2.5 provides concluding remarks.

2.2   Relational Data

It is assumed that we are given relational data that resides in a relational database; however, the framework can also be defined for data stored in a deductive database.

Definition 2.1 (*Relational database*) A relational database can be defined in the context of MRDM by the following notions.

- A relation schema is an expression of the form $R(a_1, a_2, \ldots, a_n)$, where R is a relation name, and a_i $(1 \leq i \leq n)$ are the attributes.
- A relation is a subset of the Cartesian product $V_{a_1} \times V_{a_2} \times \cdots \times V_{a_n}$, where V_{a_i} $(1 \leq i \leq n)$ are the value sets of attributes a_i.
- A relational database $D = T \cup B$ is a collection of logically connected relations, where $T = \{R_1^T, R_2^T, \ldots, R_{n_T}^T\}$ and $B = \{R_1^B, R_2^B, \ldots, R_{n_B}^B\}$ consist of target and background relations, respectively.

The target table (i.e. relation[1]) includes objects to be analyzed, e.g. objects for which association rules are mined. Such objects may reside in more than one table; for example, each target table includes the objects of one class. Background tables include additional objects which are directly or indirectly joined to the objects of the target table. The same terms are used for the objects of the target and background tables, i.e. the target and background objects.

Example 2.1 Given a database $D = \{customer\} \cup \{product, purchase\}$ for the customers of a grocery store.

customer

id	name	age	gender	income	class
1	Adam Smith	36	male	1500	yes
2	Tina Jackson	33	female	2500	yes
3	Ann Thompson	30	female	1800	no
4	Susan Clark	30	female	1800	yes
5	Eve Smith	26	female	2500	yes
6	John Clark	29	male	3000	yes
7	Jack Thompson	33	male	1800	no

married_to

id	cust_id$_1$	cust_id$_2$
1	5	1
2	6	4
3	3	7

---

[1] The notions of relation and table are used in this monograph interchangeably.

purchase					product		
id	cust$_{id}$	prod$_{id}$	amount	date	id	name	price
1	1	1	1	24/06	1	bread	2.00
2	1	3	2	24/06	2	butter	3.50
3	2	1	1	25/06	3	milk	2.50
4	2	3	1	26/06	4	tea	5.00
5	4	6	1	26/06	5	coffee	6.00
6	4	2	3	26/06	6	cigarettes	12.00
7	6	5	3	27/06			
8	3	4	1	27/06			

The target table *customer* includes basic data about customers. The data is divided into two groups according to the values of the attribute *class*. The background tables include information on marriage couples (*married_to*) and that on products purchased by the customers (*product* and *purchase*).

To consider objects apart from the tables they belong to, the notion of relational object is used.

Definition 2.2 (*Relational object*) Given a database relation with the schema $R(a_1, a_2, \ldots, a_n)$. An expression of the form $R(v_1, v_2, \ldots, v_n)$ is an object of R if and only if $(v_1, v_2, \ldots, v_n)$ is a tuple of R.

For example, the first tuple of table *customer* from Example 2.1 is represented by the object *customer*(1, *Adam Smith*, 36, *male*, 1500, *yes*).

A relational database is represented by an information system that is constructed based on the standard information system [71].[2]

Definition 2.3 (*Information system*) An information system is a pair $IS = (U, A)$, where U is a non-empty finite set of objects, called the universe, and A is a non-empty finite set of attributes.

The information system for storing relational data is constructed as follows. Consider a database $D = T \cup B$. Let $U_{D_T} = T$, $U_{D_B} = B$, $A_{D_T} = \bigcup_{R \in T} A_R$,[3] and $A_{D_B} = \bigcup_{R \in B} A_R$.

Definition 2.4 (*Information system for a relational database*) A relational database $D = T \cup B$ is represented by an information system $IS_D = (U_D, A_D)$, where

- $U_D = U_{D_T} \cup U_{D_B}$ is a non-empty finite set of objects, called the universe,
- $A_D = A_{D_T} \cup A_{D_B}$ is a non-empty finite set of attributes.

---

[2]The standard information system is understood as the Pawlak information system.

[3]A_R denotes here the set of all attributes of relation R.

Example 2.2 Database D of Example 2.1 can be represented by information system $IS_D = (U_D, A_D)$, where $U_D = U_{D_T} \cup U_{D_B}, A_D = A_{D_T} \cup A_{D_B}$ are defined as follows:
U_{D_T} = {$customer(1, Adam\ Smith, 36, male, 1500, yes), \ldots, customer(7, Jack\ Thompson, 33, male, 1800, no)$},
$U_{D_B} = \{married\_to(1, 5, 1) \ldots, married\_to(3, 3, 7), purchase(1, 1, 1, 1, 24/06), \ldots, purchase(8, 3, 4, 1, 27/06), product(1, bread, 2.00), \ldots, product(6, cigarettes, 12.00)\}$,
A_{D_T} = {$customer.id, customer.name, customer.age, customer.gender, customer.income, customer.class$},
A_{D_B} = {$married\_to.id, married\_to.cust\_id_1, married\_to.cust\_id_2, purchase.id, purchase.cust\_id, purchase.prod\_id, purchase.amount, purchase.date, product.id, product.name, product.price$}.[4]

2.3   Relational Information

Essential information acquired from relational data is expressed by descriptions of target objects. The descriptions are used in a sense to identify the objects, i.e. the objects are compared to each other or to patterns (e.g. classification rules) based on their descriptions. For each target object its description is constructed based on background relations. To construct such descriptions, the notion of related set is introduced [36].

Definition 2.5 (*Related objects*) Object o is related to object o', denoted by $o \sim o'$, if and only if there exists a key attribute joining o with o'.[5]

In this approach, the key attribute is, in general, understood as an important attribute for joining tables. It is usually a primary or foreign key. However, in some cases, it can also be another attribute by which one table can be joined with another table or with itself.

A target object description is expressed by a set of background objects joined with the target object. More precisely.

Definition 2.6 (*Related set*) A related set of a target object o, denoted by $rlt(o)$, is a set of background objects directly or indirectly related to the target object.

Each target object in this approach is processed along with its related set.

Example 2.3 Consider the target objects $o_1 = customer(1, Adam\ Smith, 36, male, 1500, yes)$, $o_2 = customer(2, Tina\ Jackson, 33, female, 2500, yes)$ from the information system of Example 2.2.

---

[4]It is assumed that the value of an attribute is specified for a given object if and only if the object belongs to the relation whose schema includes the attribute.

[5]The tables the objects belong to are not assumed to be different.

The related sets of o_1 and o_2 are $rlt(o_1) = \{married\_to(1, 5, 1), purchase(1, 1,$ $1, 1, 24/06), purchase(2, 1, 3, 2, 24/06), product(1, bread, 2.00), product(3, milk,$ $2.50)\}$ and $rlt(o_2) = \{purchase(3, 2, 1, 1, 25/06), purchase(4, 2, 3, 1, 26/06),$ $product(1, bread, 2.00), product(3, milk, 2.50)\}$, respectively.
The objects of relation $purchase$ ($product$) are directly (indirectly) related to the target objects by attribute c_id (by relation $purchase$ and attribute p_id).

For a given target object one can usually obtain more than one description, each of which describes the object with different precision. The objective is to choose an appropriate description of the target object with respect to a given data mining task. The precision of the target object description (i.e. the related set) can be tuned by its depth level. To define a related set of a given depth level, Definition 2.5 is generalized.

Definition 2.7 (*n-related objects*) Object o_0 is n-related to object o_n, denoted by $o_0 \overset{n}{\sim} o_n$, if and only if there exists o_{i+1} such that $o_i \sim o_{i+1}$, where $n > 0$ and $0 \leq i \leq n - 1$.

One can note that for $n = 1$ Definitions 2.5 and 2.7 are equivalent.
A related set of a given depth level is defined as follows.

Definition 2.8 (*n-related set*) The nth depth level related set of a target object o, denoted by $rlt^n(o)$, is a set of background objects, each of which are m-related to object o and $m \leq n$.

It is assumed that for each $o \in U_{D_T}$ we have $rlt^0(o) = \emptyset$. It is reasonable to consider a target object without its related set (i.e. the related set is empty) when the object itself includes information, i.e. descriptive attributes occur in the target relation (e.g. attribute *class* in relation *customer*).

Example 2.4 Consider the target object o_2 from Example 2.3.
We can obtain two different non-empty descriptions of o, namely $rlt^1(o_2) = \{purchase(3, 2, 1, 1, 25/06), purchase(4, 2, 3, 1, 26/06)\}$ and $rlt^2(o_2) = rlt(o_2)$.

A target object with its related sets can be presented in the form of a graph.

Definition 2.9 (*Directed graph of related set*) Given a target object o. Let $dl(o')$ be the depth level of an object $o' \in \{o\} \cup rlt(o)$. A target object o with its related set $rlt(o)$ can be presented in the form of the directed graph $G_o = (V, E)$ where $V = \{o\} \cup rlt(o)$ and $E = \{(o', o'') \in V \times V : o' \sim o'', dl(o') < dl(o'')\}$.

The directed graph illustrates how a related set of a given target object is formed (see Fig. 2.1). If the way the object description is formed is not essential, a target object with its related set can be presented using an undirected graph.

Definition 2.10 (*Undirected graph of related set*) A target object o with its related set $rlt(o)$ can be presented in the form of the undirected graph $G_o = (V, E)$ where $V = \{o\} \cup rlt(o)$ and $E = \{\{o', o''\} \subseteq V : o' \sim o''\}$.

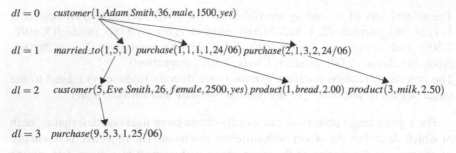

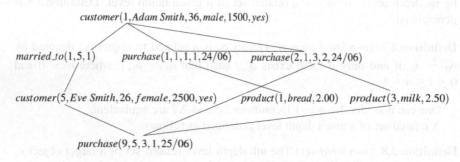

Fig. 2.1 Directed graph for the first customer from Example 2.1 (For illustrative purposes table *purchase* is extended by the tuple (9, 5, 3, 1, 25/06).)

Fig. 2.2 Undirected graph for the first customer from Example 2.1

An undirected graph enables to check if two object are *n*-related.

Proposition 2.1 *Given a target object o and its related set rlt(o). Objects o′ and o″ such that o′, o″ ∈ {o} ∪ rlt(o) are n-related if and only if there exists in G_o a path of length n joining o′ and o″.*

As it can be observed in Fig. 2.2, two objects can be related in more than one way. For example, objects *customer*(1, *Adam Smith*, 36, *male*, 1500, *yes*) and *purchase*(2, 1, 3, 2, 24/06) are 1-related and 2-related.

A related set of a given target object can be viewed as its specific description. In order to derive relational patterns the target object description is generalized. To obtain a general (i.e. abstract) description of a target object itself and its related set, they both are generalized.

Definition 2.11 (*Generalized target object*) A generalized target object o, denoted by o_{gen}, is the target object with certain components replaced according to a given substitution.[6]

---

[6]A component of an object can be replaced with either a variable, a set of constants, or symbol "_" if the component is not important for the consideration.

Definition 2.12 (*Generalized related set*) A generalized related set of a target object o, denoted by $rlt_{gen}(o)$, is the related set with certain components replaced according to the substitution (partially) constructed during generalization of the target object.

A generalized n-related set is defined in an analogous way.

Related sets can be generalized in a variety of ways (for more details see [36]). A method for generalization can be developed taking into consideration a language bias.

Example 2.5 Consider again the target object $o = customer(2, Tina\ Jackson, 33, female, 2500, yes)$ from Example 2.3 and its related set $rlt^2(o) = \{purchase(3, 2, 1, 1, 25/06), purchase(4, 2, 3, 1, 26/06), product(1, bread, 2.00), product(3, milk, 2.50)\}$.

The generalized target object and its related set can be of the following forms $o_{gen} = customer(A, \_, \_, \_, \_, yes)$ and $rlt^2_{gen}(o) = \{purchase(B, A, C, \_, \_), product(C, \{bread, milk\}, \_)\}$,[7] respectively.

An object of the relation *customer* can be generalized according to the following language bias constraint $mode(customer(+type(c\_id), \_, \_, \_, \#[yes, no]))$, which means that the first argument of the relation *customer* has to be replaced with an input variable of a type that is the same as that of attribute c_id, the last one can be replaced with *yes* or *no* (i.e. the class label), and the remaining arguments are omitted. Object o is generalized according to the substitution $\{2/A, Tina\ Jackson/\_, 33/\_, female/\_, income/\_\}$.

As presented above, each target object is represented by the set of background objects related to the target object. It is natural to treat such a set as *a granule of objects drawn together by their relationships with the target object*. Therefore, we consider a granule defined by the pair $(o, rlt(o))$, where o is a target object from a given information system.

For generalized related sets, information granules are defined by their syntax and semantics. For this purpose, the method for constructing information granules [83] is extended to a relational case.

In the approach, an elementary granule is defined by a conjunction of relational descriptors, i.e. expressions of the form $R(t_1, t_2, \ldots, t_n)$, where R is a relation name, and t_i $(1 \leq i \leq n)$ are the terms (constants or variables).

Given information system $IS_D = (U_D, A_D)$.

- A generalized target object o_{gen} of object o from IS_D is a trivial elementary granule, i.e. a single relational descriptor.

  The meaning (i.e. semantics) of the granule, denoted by $SEM_{IS_D}(o_{gen})$, is the set of target objects that satisfy the descriptor.

---

[7] The denotation $\{v_1, v_2, \ldots, v_n\}$ that occurs in an object argument list means that the corresponding attribute may take any of the values $v_1, v_2, \ldots, v_n$. We assume that sets are formed for attributes that take on a relatively small number of values. Otherwise, the attributes are previously discretized.

- A generalized related set $rlt_{gen}(o)$ of target object o from IS_D is an elementary granule where each descriptor is constructed based on a background relation. The meaning of the granule, denoted by $SEM_{IS_D}(rlt_{gen}(o))$, is the set of target objects for each of which there exists a substitution such that each descriptor under the substitution is satisfied.
- A generalized target object o_{gen} with its generalized related set $rlt_{gen}(o)$ is represented by the granule $(o_{gen}, rlt_{gen}(o))$.
  The meaning of the granule is $SEM_{IS_D}\left((o_{gen}, rlt_{gen}(o))\right) = (SEM_{IS_D}(o_{gen}), SEM_{IS_D}(rlt_{gen}(o)))$.

Example 2.6 Consider the generalized target object from Example 2.5: $o_{gen} = customer(A, \_, \_, \_, \_, yes)$ and $rlt_{gen}^2(o) = \{purchase(B, A, C, \_, \_), product(C, \{bread, milk\}, \_)\}$.
The meaning of the granule $(o_{gen}, rlt_{gen}^2(o))$ is $SEM_{IS_D}\left((o_{gen}, rlt_{gen}^2(o))\right) = (\{o_1, o_2, o_4, o_5, o_6\}, \{o_1, o_2\})$ (o_i stands for the i-th customer of database D).

Information granules defined as above can be viewed as an abstract representation of relational data. The accuracy level of the representation can easily be changed by taking other depth level of related sets. Furthermore, a representation constructed based on the information granules obtained for all target objects is treated in the approach as the search space for discovering patterns. Thanks to this, the size of the search space may significantly be limited.

A granularity of the universe is defined by the set $\{SEM_{IS_D}(rlt_{gen}^n(o)) : o \in U_{D_T}\}$. Thus different depth levels of related sets correspond to different levels of information granulation. As the depth level increases, a lower-level granularity is obtained.

2.4 Relational Knowledge

The information granules defined in the previous section are the basis for the discovery of relational knowledge. Thanks to constructing such granules we are able to obtain knowledge of different types. Therefore, we can consider as granules, e.g. frequent patterns and relational association rules, relational classification rules, and relational clusters and their descriptions.

Firstly, basic definitions will be restated (cf. [25]).

Definition 2.13 (*Relational pattern*) A relational pattern is an expression of the form[8]

$$R_1(t_1^1, t_2^1, \ldots, t_{n_1}^1) \wedge R_2(t_1^2, t_2^2, \ldots, t_{n_2}^2) \wedge \cdots \wedge R_m(t_1^m, t_2^m, \ldots, t_{n_m}^m),$$

where R_i ($1 \leq i \leq m$) are relations, and indexed t are the terms (constants or variables).

---

[8]One of relations R_i is usually considered as the target one. However, such a relation, as in this approach, may be determined externally, i.e. it occurs in the database but not in the pattern.

For simplicity's sake we denote a relational pattern as α.

The frequency of a pattern α is the ratio between the number of objects that satisfy α and the number of all objects under consideration.

Definition 2.14 (*Relational frequent pattern*) A relational frequent pattern is a relational pattern that occurs in a given database with the frequency not less than a given threshold.

Definition 2.15 (*Relational association rule*) An association rule is an expression of the form $\alpha \rightarrow \beta$, where α and β are relational (frequent) patterns and α is more general than β.

The frequency of an association rule $\alpha \rightarrow \beta$ is the frequency of β. The confidence of association rule $\alpha \rightarrow \beta$ is the ratio between the frequency of β and that of α.

Definition 2.16 (*Relational classification rule*) A relational classification rule is an expression of the form[9]

$$R(t_1, t_2, \ldots, t_n) \leftarrow R_1(t_1^1, t_2^1, \ldots, t_{n_1}^1) \wedge R_2(t_1^2, t_2^2, \ldots, t_{n_2}^2) \wedge \cdots \wedge R_m(t_1^m, t_2^m, \ldots, t_{n_m}^m),$$

where R is a target relation, R_i ($1 \leq i \leq m$) are background relations, and indexed t are terms.

For simplicity, a relational classification rule is denoted as $\alpha \leftarrow \beta$.

The accuracy (coverage) of the rule $\alpha \leftarrow \beta$ is the ratio between the number of objects that satisfy $\alpha \wedge \beta$ and the number of objects that satisfy β (α).

Example 2.7 Assume that we discover associations involving the customers from database D of Example 2.1. Table *customer* is therefore the target one, however the division into classes is not taken into account.

Given patterns $\alpha = customer(A, \_, \_, \_, \_, \_) \wedge purchase(B, A, C, \_, \_)$ and $\beta = \alpha \wedge product(C, \{bread, milk\}, \_)\}$. Patterns α and β are satisfied by objects o_1, o_2, o_3, o_4, o_6 and o_1, o_2, respectively. Hence, the frequencies of α and β are 5/7 and 2/7, respectively.

Since α is more general than β we can build the following association rule $\alpha \rightarrow \beta$. The frequency and confidence of $\alpha \rightarrow \beta$ are 2/7 and 2/5, respectively.

Consider information system $IS_D = (U_D, A_D)$. Relational patterns are represented by granules as follows.

- A relational (frequent) pattern α in IS_D is represented by the granule $(o_{gen}, rlt_{gen}(o))$. The meaning of the granule is $SEM_{IS_D}(\alpha) = (SEM_{IS_D}(o_{gen}), SEM_{IS_D}(rlt_{gen}(o)))$.

  The pattern's frequency can be calculated by $freq_{IS_D}(\alpha) = \frac{card(SEM_{IS_D}(rlt_{gen}(o)))}{card(SEM_{IS_D}(o_{gen}))}$.

---

[9]One can also consider rules including negated descriptors or conditions formed based on arguments of descriptors previously added.

- A set of relational (frequent) patterns is represented by the set of granules $\{\alpha_i : 1 \leq i \leq k\}$, where k is the cardinality of the set of rules.
  The meaning of the granule is $\{SEM_{IS_D}(\alpha_i) : 1 \leq i \leq k\}$.
- A relational association rule $\alpha \rightarrow \beta$ in IS_D is represented by the granule (α, β), where α and β are defined, respectively, by $(o_{gen}, rlt'_{gen}(o))$ and $(o_{gen}, rlt_{gen}(o))$ such that $SEM_{IS_D}(rlt_{gen}(o)) \subseteq SEM_{IS_D}(rlt'_{gen}(o))$.
  The meaning of the granule is $SIM_{IS_D}((\alpha, \beta)) = (SIM_{IS_D}(\alpha), SIM_{IS_D}(\beta))$.
  Since any association rule is constructed based on patterns that are discovered over the same relation (i.e. both patterns are checked to be satisfied for objects of the same relation), the meaning of the granule can be written in a simpler form, that is, $SIM_{IS_D}((\alpha, \beta)) = (SEM_{IS_D}(o_{gen}), SEM_{IS_D}(rlt'_{gen}(o)), SEM_{IS_D}(rlt_{gen}(o)))$.
  The rule's frequency and confidence can be calculated by $freq_{IS_D}(\alpha \rightarrow \beta) = freq_{IS_D}(\beta)$ and $conf_{IS_D}(\alpha \rightarrow \beta) = \frac{freq_{IS_D}(\beta)}{freq_{IS_D}(\alpha)}$, respectively.
- A set of relational association rules is represented by the set of granules $\{(\alpha_i, \beta_i) : 1 \leq i \leq k\}$, where k is the cardinality of the set of rules.
  The meaning of the granule is $\{SEM_{IS_D}((\alpha_i, \beta_i)) : 1 \leq i \leq k\}$.
- A relational classification rule $\alpha \leftarrow \beta$ in IS_D is represented by the granule (α, β), where α and β correspond to o_{gen} and $rlt_{gen}(o)$, respectively.
  The meaning of the granule is $SIM_{IS_D}((\alpha, \beta)) = (SIM_{IS_D}(\alpha), SIM_{IS_D}(\beta))$.
  The rule's accuracy and coverage can be computed by $acc_{IS_D}(\alpha \leftarrow \beta) = \frac{|SEM_{IS_D}(o_{gen}) \cap SEM_{IS_D}(rlt_{gen}(o))|}{|SEM_{IS_D}(rlt_{gen}(o))|}$ and $cov_{IS_D}(\alpha \leftarrow \beta) = \frac{|SEM_{IS_D}(o_{gen}) \cap SEM_{IS_D}(rlt_{gen}(o))|}{|SEM_{IS_D}(o_{gen})|}$, respectively.
- A set of relational classification rules is represented by the set of granules $\{(\alpha_i, \beta_i) : 1 \leq i \leq k\}$, where k is the cardinality of the set of rules.
  The meaning of the granule is $\{SEM_{IS_D}((\alpha_i, \beta_i)) : 1 \leq i \leq k\}$.

Example 2.8 Given information system IS_D from Example 2.2 and patterns $\alpha = customer(A, \_, \_, \_, \_, \_) \wedge purchase(B, A, C, \_, \_)$ and $\beta = \alpha \wedge product(C, \{bread, milk\}, \_)$.
Consider the following generalizations of the object $o = customer(2, Tina Jackson, 33, female, 2500, yes)$: $o_{gen} = customer(A, \_, \_, \_, \_, \_)$, $rlt^1_{gen}(o) = \{purchase(B, A, C, \_, \_)\}$, $rlt^2_{gen}(o) = \{purchase(B, A, C, \_, \_), product(C, \{bread, milk\}, \_)\}$.
Patterns α and β can be represented, respectively, by granules $(o_{gen}, rlt^1_{gen}(o))$ and $(o_{gen}, rlt^2_{gen}(o))$ with the meanings $SEM_{IS_D}(\alpha) = (\{o_1, \ldots, o_7\}, \{o_1, o_2, o_3, o_4, o_6\})$ and $SEM_{IS_D}(\beta) = (\{o_1, \ldots, o_7\}, \{o_1, o_2\})$.
The frequencies of α and β are $freq_{IS_D}(\alpha) = \frac{|SEM_{IS_D}(rlt^1_{gen}(o))|}{|SEM_{IS_D}(o_{gen})|} = 5/7$ and $freq_{IS_D}(\beta) = \frac{|SEM_{IS_D}(rlt^2_{gen}(o))|}{|SEM_{IS_D}(o_{gen})|} = 2/7$.
Consider also the association rule $\alpha \rightarrow \beta$. The meaning of the rule is $SEM_{IS_D}(\alpha \rightarrow \beta) = (\{o_1, \ldots, o_7\}, \{o_1, o_2, o_3, o_4, o_6\}, \{o_1, o_2\})$. The frequency and confidence of $\alpha \rightarrow \beta$ are $freq_{IS_D}(\alpha \rightarrow \beta) = freq_{IS_D}(\beta) = 2/7$ and $conf_{IS_D}(\alpha \rightarrow \beta) = \frac{freq_{IS_D}(\beta)}{freq_{IS_D}(\alpha)} = 2/5$.

2.5 Conclusions

This chapter has introduced a general framework for mining relational data. The structure for storing relational data in this framework is an information system that is constructed by adapting the notion of the standard information system. Information granules derived from the information system are used to construct relational patterns such as frequent patterns, association rules, and classification rules.

The introduced framework can be summarized as follows.

1. The framework can be helpful when a given database consists of many tables and some background objects are joined with the target ones through a number of tables. In this case, there arises the problem of how deeply one should search the database for background objects that are joined with the target ones. In the framework the search level can easily be changed so as to adjust the target object representation to a given data mining task.
2. The framework can also be useful when the search space limitation achieved by a language bias is not sufficient. The search space can additionally be limited since this is given as a set of information granules derived from the data.
3. The framework has an advantage over the ILP and RDB frameworks in terms of generation of patterns. Namely, the framework, unlike others, partially unifies the process of discovering patterns from data. This is done by constructing the search space based on information granules. The patterns can thus directly be obtained from such granules or constructed based on them.

Chapter 3
Properties of Granular-Relational Data Mining Framework

3.1 Introduction

Mining relation data is a more complicated and complex task than in the case of data stored in a single table. Relational data is distributed over multiple tables, which causes that the issues such as relational objects representation (1), search space limitation (2), and relational patterns generation (3) need more attention [44].

1. An object of a single table database is represented by a tuple of table attributes values. An object of a database with a relational structure can be represented not only by a tuple that belongs to a table to be analyzed, but also by tuples of other tables that are directly or indirectly joined to the table under consideration. Therefore, relational objects representation can vary, depending on given representation precision and data mining task.
2. Due to multiple tables the search space for discovering relational patterns may be very huge. This problem is typically overcome by applying a language bias. It imposes some constraints on patterns to be discovered, thereby the search space is limited. However, the search space after such a limitation has been imposed on it can still be large.
3. A method for deriving patterns from data is usually provided not by a given framework for mining relational data, but by a concrete algorithm that can be defined in the framework. Therefore, the whole process of the generation of patterns may be conducted from scratch at each time when any algorithm parameter changes.

The goal of this chapter is to address the three above problems. To tackle them, properties of the granular computing framework for mining relational data introduced in Chap. 3 are investigated. The crucial issue in the three tasks is to find a proper depth level of searching the data. It can be done globally, i.e. the same level for all the

© Springer International Publishing AG 2017

P. Hońko, *Granular-Relational Data Mining*, Studies in Computational Intelligence 702, DOI 10.1007/978-3-319-52751-2_3 tasks, or locally, i.e. the level is individually defined for each task.[1] The properties investigated in this work can facilitate the process of finding a proper depth level for any of the three tasks.

The remaining of the chapter is organized as follows. Sections 3.2, 3.3, and 3.4 investigate, respectively, properties of relational objects representation, properties that can be useful for limiting the search space, and properties that can improve the process of relational pattern generation. Section 3.5 provides concluding remarks.

3.2 Relational Objects Representation

As mentioned in Sect. 3.1 an object can be represented not only by the tuple that belongs to one relation (i.e. the target relation), but also by tuples that belong to other relations that are directly or indirectly joined to the target one. The crucial task is therefore to find a representation proper in terms of generality. That is, on the one hand, it is specific enough to identify objects, and on the other hand, it is general enough to avoid too detailed information.

Traditional algorithms for relational data mining do not use explicit objects representation; however, they are able to use a depth level (usually specified by an expert) during the construction of patterns.[2] It means that all background objects of the given depth level are considered as the representation of the target ones. Therefore, any time target objects are processed, the background ones are scanned up to a give depth level.

Considering explicit objects representations can shorten the process of pattern generation since unessential features of relational objects can be removed during the construction of the representations.

The remaining part of this subsection shows properties of the framework that can be useful for constructing relational objects representation.

Consider a finite database $D = T \cup B$, where $T = U_{D_T}$ and $B = U_{D_B}$ are, respectively, the sets of target and background relations of D. Let $IS_D = (U_D, A_D)$ be an information system of a given database D, where $U_D = U_{D_T} \cup U_{D_B}$. Let also n_D denote the maximal depth level of D.

Firstly, a simple definition is introduced to compare object descriptions in terms of generality. Here a description of an object is defined by its related set.

Definition 3.1 (Generality relation of related sets) Let $rlt(o)$ and $rlt'(o)$ be descriptions of an object $o \in U_{D_T}$. If $rlt(o) \subseteq rlt'(o)$ than $rlt(o)$ is more general than or equal to $rlt'(o)$ (equivalently, $rlt'(o)$ is more specific than or equal to $rlt(o)$).

Let us start with two obvious properties.

---

[1]If the depth level for objects representation is i, then those for the remaining tasks may be not higher than i.

[2]Learning from interpretations [19] is an alternative way of handling relational data. Interpretations correspond to non-abstract objects representations introduced in Chap. 2.

Proposition 3.1 [3] $\displaystyle \bigvee_{o \in U_{D_T}} rlt(o) \subseteq U_{D_B}$.

Proposition 3.2 $\displaystyle \bigvee_{o \in U_{D_T}} rlt^i(o) \subseteq rlt^{i+j}(o)$, where $0 \leq i \leq n_D, 0 \leq j \leq n_D - i$.

Based on the above property, we can see that with increasing depth level, object description (i.e. its related set) may become more specific.

Proposition 3.3 $\displaystyle \bigvee_{o \in U_{D_T}} \bigvee_{0 \leq i < n_D} \left[rlt^i(o) = rlt^{i+1}(o) \Rightarrow \bigvee_{1 < j \leq n_D - i} rlt^i(o) = rlt^{i+j}(o) \right]$.

This property enables us to find the deepest level of object description, which may be lower than the maximal depth level of the database.

Let $S(o) = \{rlt^i(o) : 0 \leq i \leq n_D, rlt^i(o) \neq \emptyset\}$, where $o \in U_{D_T}$.

By Proposition 3.3 we obtain the following property.

Proposition 3.4 *The following holds:* $\displaystyle \bigvee_{o \in U_{D_T}} |S(o)| \leq n_D$.

Proposition 3.5 *For any $o \in U_{D_T}$ the set $S(o)$ is well totally ordered by the relation $\subseteq$.*

This property enable us to order all object descriptions according to generality. Furthermore, this order can be known a priori (see Proposition 3.2).

The set $S(o)$ is well ordered, hence when considering any number of object descriptions, we can always find the most general one. Since any subset of the set $S(o)$ is finite, we can also find the most specific object description.

Let $S^i = \{rlt^i(o) : o \in U_{D_T}, rlt^i(o) \neq \emptyset\}$.

Proposition 3.6 *The following hold:*

1. $\bigcup S^i = \emptyset$ for $i = 0$;
2. S^i is a partial covering of set U_{D_B} for $1 \leq i < n_D$;
3. S^i is a total covering of set U_{D_B} for $i = n_D$.

Remark 3.1 The set S^i can be treated as the i-th level description of all target objects (target relation description for short).

Definition 3.2 (Relation $\preceq$ on $\mathscr{S}$) Let $\mathscr{S} = \{S^i : 0 \leq i \leq n_D\}$. We define a relation $\preceq$ on $\mathscr{S}$ as follows:

$$S^i \preceq S^j \Leftrightarrow \bigvee_{X \subseteq S^i} \exists_{Y \in S^j} X \subseteq Y.$$

If $S^i \preceq S^j$, we say that S^i is more general than or equal to S^j (equivalently, S^j is more specific than or equal to S^i).

Proposition 3.7 $S^i \preceq S^{i+j}$, where $0 \leq i \leq n_D, 0 \leq j \leq n_D - i$.

---

[3] Proofs of the propositions formulated in this chapter can be found in [44].

Based on the above property, we can see that with increasing depth level, target relation description may become more specific.

Proposition 3.8 *The set $\mathscr{S}$ is well weakly ordered by the relation $\preceq$.*

Since the set $\mathscr{S}$ is weakly ordered, we can compare in terms of generality any two target relation descriptions. Despite the antisymmetry does not hold, the set $\mathscr{S}$ can be totally ordered based on the depth level according to Proposition 3.7.

The set $\mathscr{S}$ is well ordered, hence when considering any number of target relation descriptions, we can always find the most general one. Since any subset of the set $\mathscr{S}$ is finite, we can also find the most specific target relation description.

3.3   Search Space Limitation

Relational data is distributed over multiple tables, thereby the search space for discovering relational patterns may be very huge. This problem is typically overcome by applying a language bias. It imposes some constraints on patterns to be discovered, thereby the search space is limited. However, the search space after such a limitation has been imposed on it may still be large.

The remaining part of this subsection shows properties of the framework thanks to which the search space can additionally be limited.

To define the search space the following set is used. Let $At_{IS_D} = \{R_i(t_1^i, \ldots, t_{n_i}^i) : t_j^i \in Tm \ (j = 1, \ldots, n_i), R_i \in rel(D)\}$ be the set of all atom formulas of information system IS_D, where Tm is the set of all terms (constants or variables), and $rel(D)$ is the set of all relation names of a database D.

Remark 3.2 The family $\mathscr{P}(At_{IS_D})$ can be treated as the unlimited search space for relational patterns.

We have the following relationship between abstract object descriptions and the search space.

Proposition 3.9 $\quad \underset{o \in U_{D_T}}{\forall} \ rlt_{gen}(o) \in \mathscr{P}(At_{IS_D}).$

3.3.1   Syntactic Comparison of Abstract Objects Descriptions

This subsection shows how abstract objects descriptions, i.e. generalized related sets, can syntactically be compared in terms of generality.[4]

---

[4]One should distinguish between the syntax of a related set and the syntactical comparison of related sets. The former concerns the form of the related set, whereas the latter does the way the relate sets are compared. Analogously for semantics.

An abstract object description, denoted by $rlt_{gen}(o)$, is, in fact, obtained by applying a substitution σ to the object description $rlt(o)$, i.e. $rlt_{gen}(o) = rlt(o)\sigma$.

Syntactic comparison of abstract objects descriptions is possible if they are constructed in the same way regardless of the depth level. Therefore, the following assumption is made.

Assumption 3.1 $\underset{1 \leq i < j \leq n_D}{\forall} \underset{o \in U_{D_T}}{\forall} rlt^i(o)\sigma_i = rlt^i(o)\sigma_j$

It is assumed above that a term to be replaced is substituted according to the same binding regardless of the depth level the substitution is defined for.

Proposition 3.10 *Under Assumption 3.1,* $\underset{o \in U_{D_T}}{\forall} rlt^i_{gen}(o) \subseteq rlt^{i+j}_{gen}(o)$, *where* $0 \leq i \leq n_D, 0 \leq j \leq n_D - i$.

Based on the above property, we can see that with increasing depth level, abstract object description may become more specific.

Let $S_{gen}(o) = \{rlt^i_{gen}(o) : 0 \leq i \leq n_D, rlt^i(o) \neq \emptyset\}$, where $o \in U_{D_T}$.

Proposition 3.11 *The set $S_{gen}(o)$ ($o \in U_{D_T}$) is partially ordered by the relation $\subseteq$.*

Thanks to this property, we can find the most (or least) general object descriptions.

Proposition 3.12 *Under Assumption 3.1, the set $S_{gen}(o)$ ($o \in U_{D_T}$) is well totally ordered by the relation $\subseteq$.*

By the above properties, we can order any subset of the set $S_{gen}(o)$ according to generality and find in the subset the most (or least) general object descriptions.

Let $S^i_{gen} = \{rlt^i_{gen}(o) : o \in U_{D_T}, rlt^i(o) \neq \emptyset\}$.

Remark 3.3 The set S^i_{gen} can be treated as the search space limited by the i-th depth level.

Let $\mathscr{S}_{gen} = \{S^i_{gen} : 1 \leq i \leq n_D\}$.

Proposition 3.13 *Under Assumption 3.1, $S^i_{gen} \preceq S^{i+j}_{gen}$, where* $0 \leq i \leq n_D, 0 \leq j \leq n_D - i$.

Based on the above property, we can see that with increasing depth level, abstract target relation description may become more specific.

Proposition 3.14 *The set $\mathscr{S}_{gen}$ is partially ordered by the relation $\preceq$.*

Thanks to this property, we can find the most (or least) general target relation descriptions.

Proposition 3.15 *Under Assumption 3.1, the set $\mathscr{S}_{gen}$ is well totally ordered by the relation $\preceq$.*

By the above properties, we can order any subset of the set $\mathscr{S}_{gen}$ according to generality and find in the subset the most (or least) general target relation descriptions.

3.3.2   Semantic Comparison of Abstract Objects Descriptions

When comparing patterns in terms of generality, a semantic order may be more important than the syntactic one. This subsection shows how abstract objects descriptions can semantically be compared in terms of generality.

Let $\sigma_{|\sigma'} = \{t/t' \in \sigma : \exists \underset{t''}{} t/t'' \in \sigma'\}$ denote a substitution σ limited by a substitution σ' to bindings that include terms to be replaced by both σ and σ'.

Assumption 3.2 $\underset{1 \leq i < j \leq n_D}{\forall}$ σ_i and $\sigma_{j|\sigma_i}$ are equivalent.[5]

Definition 3.3 (Relation $\preceq_\theta$ on $S_{gen}(o)$) Let $S_{gen}(o)$ ($o \in U_{D_T}$) be defined as previously. A relation $\preceq_\theta$ on $S_{gen}(o)$ is defined as follows:

$$X \preceq_\theta Y \Leftrightarrow \exists_\theta X\theta \subseteq Y,$$

where $X, Y \in S_{gen}(o)$ and θ is a substitution.[6]
If $X \preceq_\theta Y$, we say that X is more general than or equal to Y (equivalently, Y is more specific than or equal to X) (cf. [75]).

Proposition 3.16 *Under Assumption 3.2,* $\underset{o \in U_{D_T}}{\forall}$ $rlt^i_{gen}(o) \preceq_\theta rlt^{i+j}_{gen}(o)$, *where* $0 \leq i \leq n_D, 0 \leq j \leq n_D - i$.

Proposition 3.17 *Under Assumption 3.1,* $rlt^i_{gen}(o) \preceq_\theta rlt^{i+j}_{gen}(o)$ *for* $\theta = \emptyset$, *where* $0 \leq i \leq n_D, 0 < j \leq n_D - i$.

Proposition 3.18 *The set* $S_{gen}(o)$ ($o \in U_{D_T}$) *is quasi ordered by the relation* $\preceq_\theta$.

Proposition 3.19 *Under Assumption 3.2, the set* $S_{gen}(o)$ ($o \in U_{D_T}$) *is well weakly ordered by the relation* $\preceq_\theta$.

Definition 3.4 (Relation $\preceq_\theta$ on $\mathscr{S}_{gen}$) Let $\mathscr{S}_{gen}$ be defined as previously. A relation $\preceq_\theta$ on $\mathscr{S}_{gen}$ is defined as follows:

$$S^i_{gen} \preceq_\theta S^j_{gen} \Leftrightarrow \underset{X \in S^i_{gen}}{\forall} \underset{Y \in S^j_{gen}}{\exists} X \preceq_\theta Y.$$

If $S^i_{gen} \preceq_\theta S^j_{gen}$, we say that S^i_{gen} is more general than or equal to S^j_{gen} (equivalently, S^j_{gen} is more specific than or equal to S^i_{gen}).

---

[5]Two substitutions are equivalent if and only if each one can be obtained from the other one by renaming variables.

[6]The notion of substitution is used in two cases: for the generalization of related sets; for the semantic comparison of generalized related sets. To better distinguish these cases, we denote the former substitution by (indexed) σ, and the latter one by (indexed) θ.

Proposition 3.20 *Under Assumption 3.2,* $\underset{o \in U_{D_T}}{\forall} S^i_{gen} \preceq_\theta S^{i+j}_{gen}$, *where* $0 \leq i \leq n_D, 0 \leq j \leq n_D - i.$

Proposition 3.21 *The set* $\mathscr{S}_{gen}$ *is quasi ordered by the relation* $\preceq_\theta$.

Proposition 3.22 *Under Assumption 3.2, the set* $\mathscr{S}_{gen}$ *is well weakly ordered by the relation* $\preceq_\theta$.

Conclusions from the properties given above are analogous to those from Sect. 3.3.1.

3.4 Relational Patterns Generation

A method for deriving patterns from data is usually provided not by a given framework for mining relational data, but by a concrete algorithm that can be defined in the framework. Therefore, the whole process of the generation of patterns may be conducted from scratch at each time when any algorithm parameter changes.

The remaining part of this subsection shows properties of the framework that can be useful for relational patterns generation.

When generating a pattern, we mainly consider its semantics to determine the pattern's quality.

For simplification, we will use the denotation $M(\bullet)$ instead of $SEM_{IS_D}(\bullet)$, where M is the abbreviation of meaning, i.e. semantics.

It is assumed that the higher number of objects satisfying a pattern candidate is, the more general the pattern candidate is.

Proposition 3.23 $\underset{o \in U_{D_T}}{\forall} M(rlt_{gen}(o)) \subseteq U_{D_T}.$

Proposition 3.24 *Under Assumption 3.1,* $\underset{o \in U_{D_T}}{\forall} M(rlt^{i+j}_{gen}(o)) \subseteq M(rlt^i_{gen}(o))$, *where* $0 \leq i \leq n_D, 0 \leq j \leq n_D - i.$

Based on the above property, we can see that with increasing depth level, the pattern candidate may became more specific.

Let $S_M(o) = \{M(rlt^i_{gen}(o)) : 0 \leq i \leq n_D, rlt^i(o) \neq \emptyset\}$, where $o \in U_{D_T}$.

Proposition 3.25 *The set* $S_M(o)$ *(*$o \in U_{D_T}$*) is partially ordered by relation* $\subseteq$.

Thanks to this property, we can find the most (or least) general pattern candidates.

Proposition 3.26 *Under Assumption 3.1, the set* $S_M(o)$ $o \in U_{D_T}$ *is well totally ordered by the relation* $\subseteq$.

By the above properties, we can order any subset of the set $S_M(o)$ according to generality and find in the subset the most (or least) general pattern candidates.

Let $S_M^i = \{M(rlt_{gen}^i(o)) : o \in U_{D_T}\}$.

Proposition 3.27 $S_M^i \preceq S_M^{i+j}$, where $0 \leq i \leq n_D, 0 \leq j \leq n_D - i$.

Based on the above property, we can see that with increasing depth level, pattern candidates set may become more specific.

Proposition 3.28 *The set $\mathscr{S}_M$ is quasi ordered by relation $\preceq$.*

Proposition 3.29 *Under Assumption 3.1, the set $\mathscr{S}_M$ is well weakly ordered by relation $\preceq$.*

Since the set $\mathscr{S}_M$ is weakly ordered, we can compare in terms of generality any two pattern candidate sets. Despite the antisymmetry does not hold, the set $\mathscr{S}_M$ can be totally ordered based on the depth level according to Proposition 3.27.

The set $\mathscr{S}_M$ is well ordered, hence when considering any number of pattern candidate sets, we can always find the most general one. Since any subset of the set $\mathscr{S}_M$ is finite, we can also find the most specific pattern candidate set.

3.5 Conclusions

This chapter has investigated properties of the granular computing framework for mining relational data. The properties have been studied in the context of the three essential relational data mining problems: relational objects representation, search space limitation, and relational patterns generation. One can observe that they are in a sense interrelated. Namely, the level of generality of relational objects representation can determine those of the search space and pattern candidates set. Therefore, thanks to the framework, one can not only separately consider one of the above issues, but also comprehensively process relational data according to a given data mining task.

Chapter 4
Association Discovery and Classification Rule Mining

4.1 Introduction

Association discovery and classification are ones of the most extensively studied tasks in the field of data mining (see, e.g. [1, 2, 8, 33–35, 92]). These issues have also been widely investigated for relational data (see, e.g. [22, 25, 59, 93, 114]). One can indicate many different relational techniques and algorithms for both tasks; however, a unified framework for them does not seem to have been introduced so far. Such a framework is needed for unifying operations that are independent of the technique or algorithm applied for processing relational data. One can indicate the following essential operations that need to be unified: relational object representation, search space limitation and generation of relational patterns. These issues will briefly be discussed.

1. An object of a single-table database is represented by a tuple of table attribute values. An object of a database with a relational structure can be represented not only by a tuple that belongs to a table to be analyzed, but also by a certain part of the tuples of other tables that are directly or indirectly joined to the table under consideration. Therefore, relational object representation can vary depending on a given data mining task.

2. The search space for discovering relational patterns may be very huge. This problem is typically overcome by applying a language bias, which imposes some constraints on the patterns to be discovered, thereby the search space is limited. However, the search space, after such a limitation has been imposed on it, may still be large.

3. Rule-based classification is one of the most common classifying methods in data mining. Classification rules can be considered as a special case of association rules. The way of deriving both types of rules from data is usually provided not by a given framework for mining relational data, but by a concrete algorithm that can be defined in the framework; therefore, the whole process of the generation of rules may be conducted from scratch each time any of the algorithm's parameters change.

© Springer International Publishing AG 2017
P. Hońko, *Granular-Relational Data Mining*, Studies in Computational
Intelligence 702, DOI 10.1007/978-3-319-52751-2_4

When a unified framework for mining (relational) data is developed, there is a need to specify it for a given data mining task. On the one hand, such a specialized framework should be tuned for one concrete task, on the other hand, the framework should make it possible to define a spectrum of algorithms for the task to be performed, both existing and new algorithms.

The goal of this chapter is to provide frameworks for association and classification rules discovery from relational data [38, 39]. The frameworks are specialized versions of the general granular computing framework for mining relational data (see Chap. 2).

The remaining of the chapter is organized as follows. Sections 4.2 and 4.3 introduce specialized frameworks for association discovery and classification rule mining, respectively. Section 4.4 evaluates the approach's complexity. Section 4.5 provides concluding remarks.

4.2   Association Discovery

The general framework for association discovery can be outlined as follows.

1. For each target object compute its related set.
2. Generalize the related sets.
3. Compute frequent patterns based on the generalized related sets.
4. Compute association rules based on the frequent patterns.

The tools for performing the first two steps are provided by the general granular computing framework described in Chaps. 2 and 3. The third step is crucial for association discovery and will be presented in this section. The last step can be, in turn, performed by applying any existing method that derives relational association rules from frequent patterns, cf. [21].

In the following approach, frequent patterns can be generated by applying a top-down or bottom-up method.

Given:

- $IS_D = (U_D, A_D)$—the information system of database D;
- n—the depth level of related sets;
- k—the number of seeds, i.e. the number of target objects to be generalized;
- $minfreq \in (0, 1]$—the frequency threshold.

Find:

- FP—a set of frequent patterns;

Steps:

1. For each target object o of IS_D compute $rlt^n(o)$.
2. Choose randomly k seed objects from the target objects.
3. Based on the chosen objects, generate initial patterns, i.e. granules of the form $(o_{gen}, rlt^m_{gen}(o))$, where

   a. *The top-down case*: $m := 0$;
   b. *The bottom-up case*: $m := n$.

4. Add to *FP* each initial pattern with frequency not less than *minfreq*.
5. $FP' := \emptyset$; For each m-level frequent pattern p from *FP* find its all allowed

   a. *The top-down case*: $FP' = FP' \cup special(p, m)$; Next $m := m + 1$;
   b. *The bottom-up case*: $FP' = FP' \cup general(p, m)$; Next $m := m - 1$.

6. Add to *FP* each frequent pattern from *FP'* if p is not equivalent to any pattern from *FP*.
7. Repeat steps 5 and 6 until

   a. *The top-down case*: all of the patterns from step 5 are not frequent or $m > n$;
   b. *The bottom-up case*: $m < 0$.

The most important points will be discussed below.
Re 3. The way target objects and their related sets are generalized in the top-down case may vary from that in the bottom-up case.
Re 3a, b.
The top-down case: Initial patterns are to be the most general ones, hence $m = 0$.
The bottom-up case: Initial patterns are to be the most specific ones, hence $m = n$.
Re 5. Each pattern from *FP* is the one of level m when the instruction from step 5 is performed for the fist time. Otherwise, patterns of level m are obtained based on the patterns of the previous level found in step 6 (the previous loop run). Furthermore, such patterns, unlike those from the first loop run, may not be frequent, therefore, we check if a given pattern is frequent before we find its specifications/generalizations. Allowed specifications or generalizations of a given pattern are understood as those patterns that can be formed according to given constraints.
Re 5a. A specialization of a pattern p is done in one of the following ways:

1. A variable that occurs in p is replaced with a set of values the variable may take, e.g. a specialization of $customer(A, \_, \_, \_, \_, B)$ is $customer(A, \_, \_, \_, \_, 1)$[1];
2. A set of values that occurs in p as a component is replaced with its non-empty subset, e.g. a specialization of $customer(A, \_, \_, \_, \_, \_) \wedge purchase(B, A, C, \{1, 2\}, \_)$ is $customer(A, \_, \_, \_, \_, \_) \wedge purchase(B, A, C, 2, \_)$;
3. p is extended by an additional condition, e.g. a specialization of $customer(A, \_, \_, \_, \_, \_)$ is $customer(A, \_, \_, \_, \_, \_) \wedge purchase(B, A, C, 2, \_)$.

Re 5b. A generalization of a pattern p is done in one of the following ways:

1. A set of values that occurs in p as a component is replaced with a new variable, e.g. a generalization of $customer(A, \_, \_, \_, \_, 1)$ is $customer(A, \_, \_, \_, \_, B)$;
2. A set of values that occurs in p as a component is replaced with its superset, e.g. a generalization of $customer(A, \_, \_, \_, \_, \_) \wedge purchase(B, A, C, 2, \_)$ is $customer(A, \_, \_, \_, \_, \_) \wedge purchase(B, A, C, \{1, 2\}, \_)$;

---

[1]If a set of values consists of one element, then the set is replaced with the element.

3. p is reduced by removing one of its conditions, e.g. a generalization of $customer(A, \_, \_, \_, \_, \_) \ \wedge \ purchase(B, A, C, \{1, 2\}, \_)$ is $customer$ $(A, \_, \_, \_, \_, \_)$.

Furthermore, the condition from steps 5 and 6 that a pattern has to be frequent may be omitted in the bottom-up case. The condition is satisfied according to the following property: Any generalization of a frequent pattern is frequent.

The approach is illustrated by the following example.

Example 4.1 Given information system IS_D from Example 2.2. We examine the top-down case and the following settings $n = 1, k = 1, minfreq = 0.3$. We use the following constraints during the construction of patterns:

1. $mode(customer(+type(customer.id), \_, \_, \_, \_, +type(class)))$,
2. $mode(customer(+type(customer.id), \_, \_, \_, \_, \#[yes, no]))$,
3. $mode(1, purchase(+type(purchase.id), +type(cust\_id), -type(prod\_id),$ $-type(amount), \_))$,[2]
4. $mode(1, purchase(+type(purchase.id), +type(cust\_id), -type(prod\_id),$ $\#[1, 2, 3], \_))$,
5. When a pattern is being specialized, its variable can be replaced with exactly one value (i.e. a set of values is not allowed).

Suppose that $o_2 = customer(2, Tina\ Jackson, 33, female, 2500, yes)$ is a randomly chosen seed object. We have $rlt^1(o_2) = \{purchase(3, 2, 1, 1, 25/06),$ $purchase(4, 2, 3, 1, 26/06)\}$. According to the above constraints we obtain the following generalizations $o_{2_{gen}} = customer(A, \_, \_, \_, \_, B), rlt^1_{gen}(o_2) = \{purchase$ $(C, A, D, E, \_)\}$.

Let $c_1 = customer(A, \_, \_, \_, \_, B), c_2 = (customer(A, \_, \_, \_, \_, yes), c_3 = customer(A, \_, \_, \_, \_, no), p_1 = purchase(C, A, D, E, \_), p_2 = purchase$ $(C, A, D, 1, \_)\}), p_3 = purchase(C, A, D, 2, \_), p_4 = purchase(C, A, D, 3, \_)$.

The case $m = 0$. We have the initial pattern $(o_{2_{gen}}, rlt^0_{gen}(o_2)) = (c_1, \emptyset)$ with frequency 1, hence $FP = \{(o_{2_{gen}}, rlt^0_{gen}(o_2))\}$.

The specifications of the pattern are the following (frequency given after the colon): $(c_2, \emptyset) : 5/7, (c_3, \emptyset) : 2/7$.

Since only the frequency of the first new pattern is higher than $minfreq$, we obtain $FP = \{(c_1, \emptyset), (c_2, \emptyset)\}$.

The case $m = 1$. Based on the patterns from FP we obtain the following patterns of level 1: $(c_1, \{p_1\}) : 5/7, (c_2, \{p_1\}) : 4/7$.

Specializations of these patterns are:
$(c_1, \{p_2\}) : 4/7, (c_2, \{p_2\}) : 3/7, (c_1, \{p_3\}) : 1/7, (c_2, \{p_3\}) : 1/7, (c_1, \{p_4\}) : 2/7, (c_2, \{p_4\}) : 2/7$.

---

[2]An argument preceded by symbol "+" ("-") has to be replaced with an input (output) variable. The first argument of function *mode* (i.e. value 1) means that the relation *purchase* can be used in the construction of a pattern at most once.

Since $m = n$, the set of frequent patterns to be returned is of the following form
$FP = \{(c_1, \emptyset), (c_2, \emptyset), (c_1, \{p_1\}), (c_2, \{p_1\}), (c_1, \{p_2\}), (c_2, \{p_2\})\}$.
One can observe that the first and third patterns of set FP are directly obtained
from the information granules, i.e. the generalizations of object o_2 (levels 0 and 1,
respectively), whereas the remaining patterns are constructed based on the granules.

4.3 Classification Rule Mining

The following framework for generating relational classification rules is introduced.
In this approach the rules can be generated by applying a top-down or bottom-up
method.

The general framework for classification rule mining is analogous to that from
Sect. 4.2.

1. For each target object compute its related set.
2. Generalize the related sets.
3. Compute classification rules based on the frequent patterns.

A more detailed solution is given below.

Given:

- $IS_D = (U_D, A_D)$—the information system of database D;
- n—the depth level of related sets;

Find:

- RS—a set of classification rules;

Steps:

1. $RS := \emptyset$;
2. For each target object o of IS_D compute $rlt^n(o)$;
3. Choose one object from the target objects;
4. Based on the chosen object, generate an initial rule, i.e. a granule of the form
   $r = (o_{gen}, rlt^m_{gen}(o))$, where

   a. *The top-down case*: $m := 0$;
   b. *The bottom-up case*: $m := n$;

5. Refine the initial rule r:

   a. Compute a set of candidate rules:
      i. *The top-down case*: $RS' := \{r\} \cup special(r, m)$; Next $m := m + 1$;
      ii. *The bottom-up case*: $RS' := \{r\} \cup general(r, m)$; Next $m := m - 1$;
   b. $r := best\_candidate(RS')$;
   c. Repeat step 5 until $stop\_criterion(r)$;

6. $RS := RS \cup \{r\}$;
7. Repeat steps 3–6 until $stop\_criterion(RS)$;

Selected steps of the above framework will be studied.

Re 3. The way of choosing target objects is defined by the algorithm to be used. One can observe that in the top-down case the choice of a target object is not important because the generalization of each target object of a given class is the same.

Re 4. The way target objects and their related sets are generalized in the top-down case may vary from that in the bottom-up case.

Re 4a, b.

The top-down case: An initial rule is to be the most general one, hence $m = 0$.

The bottom-up case: An initial rule is to be the most specific one, hence $m = n$.

Re 5a. The function $special(r, m)$ $(general(r, m))$ returns a set of allowed specializations (generalizations) of a rule r at a level m.[3] The function works analogously to that for specializing (generalizing) frequent patterns.

Re 5b. The function $best\_candidate(S)$ returns a rule from S that has the highest quality based on a given quality measure.

Re 5c. The stop criterion is defined by a given technique or algorithm for generating classification rules. For step 7 it is done analogously.

Rule generation is illustrated by the following example.

Example 4.2 We are given information system IS_D from Example 2.2. We examine the top-down case and $n = 1$. We evaluate a rule based on its accuracy and use the following constraints during the construction of the rule:

1. $mode(customer(+type(customer.id), \_, \_, \_, \_, \#[yes, no])$,
2. $mode(1, purchase(+type(purchase.id), +type(cust\_id), -type(prod\_id), -type(amount), \_)))$,
3. $mode(1, purchase(+type(purchase.id), +type(cust\_id), -type(prod\_id), \#[1, 2, 3], \_))$,

We have $RS := \emptyset$;

Suppose that $o_2 = customer(2, Tina\ Jackson, 33, female, 2500, yes)$ is a chosen object. We have $rlt^1(o_2) = \{purchase(3, 2, 1, 1, 25/06), purchase(4, 2, 3, 1, 26/06)\}$. According to the above constraints, we get the following generalizations $o_{2_{gen}} = customer(A, \_, \_, \_, \_, B)$, $rlt^1_{gen}(o_2) = \{purchase(C, A, D, E, \_)\}$.

Let $c_1 = customer(A, \_, \_, \_, \_, yes)$, $p_1 = purchase(B, A, C, E, \_)$, $p_2 = purchase(B, A, C, 1, \_)$, $p_3 = purchase(B, A, C, 2, \_)$, $p_4 = purchase(B, A, C, 3, \_)$, $p_5 = purchase(B, A, C, \{1, 2\}, \_)$, $p_6 = purchase(B, A, C, \{1, 3\}, \_)$, $p_7 = purchase(B, A, C, \{2, 3\}, \_)$.

The case $m = 0$. We have the initial rule $r = (o_{2_{gen}}, rlt^0_{gen}(o_2)) = (c_1, \emptyset)$[4] with accuracy 5/7.

---

[3] Allowed specializations or generalizations of a given rule are understood as those rules that can be formed according to given constraints.

[4] The granule $(customer(A, \_, \_, \_, \_, yes), \emptyset)$ can be transformed into the rule $customer(A, \_, \_, \_, \_, 1) \leftarrow 1$, where the rule premise is satisfied by any object.

Since the set of specifications of r is empty under the given constraints, then r is the best candidate.

The case $m = 1$. Based on r we obtain the following rule of level 1: $r := (c_1, p_1)$.

We have the following specialization of r at level 1 (accuracy given after the colon): $(c_1, \{p_2\})$: $3/4, (c_1, \{p_3\})$: $1/1, (c_1, \{p_4\})$: $2/2, (c_1, \{p_5\})$: $3/4, (c_1, \{p_6\})$: $4/5, (c_1, \{p_7\}) : 3/3$.

We have three rules with the maximal accuracy, then the rule with the highest coverage is taken, i.e. $r := (customer(A, \_, \_, \_, \_, 1), \{purchase(A, C, \{2, 3\}, \_)\})$.

Since $m = n$, then $RS := RS \cup \{r\}$.

4.4 The Approach's Complexity

This subsection provides an analysis of the framework's time complexity. Operations such as granule formation and pattern generation (association and classification rules) are studied.

Let $n = |U_{D_T}|$ and $m = |U_{D_B}|$.

1. The cost of the formation of granules $(o, rlt(o))$ for all $o \in D_T$ is

$$T(n, m) = nm' \leq nm = O(nm),$$

   where m' is the number of all objects from the database's tables to be scanned. In a pessimistic case, we have $m' = m$.

2. The cost of the generalization of all granules $(o, rlt(o)) \in U$ is

$$T(n, m) = |U| \sum_{o' \in \{o\} \cup rlt(o)} \sum_{a \in attr(o')} 1 = n(|rlt(o)| + 1)C \leq n(m + 1)C = O(nm),$$

   where $C = \sum_{a \in attr(o')} 1$ is the cost of the generalization of an object o'.[5] C does not depend on the data size.

   Relational data is represented by a class of granules of the form $(o_{gen}, rlt_{gen}(o))$. One can observe that the size of this representation only depends on the size of U_{D_T}. Namely, we assume that a given database is representative, i.e. (almost) all relationships occur in the database. Therefore, adding new background objects does not affect (or hardly affects) the form of the generalized related sets. Hence, we can ignore the size of $rlt_{gen}(o)$ when analyzing the complexity of the approach.

1. Construct a pattern (without checking the pattern's satisfaction).
   Let o be a target object based on which a pattern is constructed. To generate a pattern we need to scan objects from $rlt_{gen}(o)$ to check which of them should be taken as the pattern's conditions. We assume that the cost of scanning a set is

---

[5]$attr(o)$ is the collection of all components of an object o.

equal to its cardinality.

The cost of pattern generation is

$$T(n) = \sum_{o' \in rlt_{gen}(o)} 1 = |rlt_{gen}(o)| \leq C = O(1),$$

where $C = \max\{|rlt_{gen}(o)| : o \in U_{D_T}\}$.

2. Compute $special(r, i)$ or $general(r, i)$ for all conditions of a pattern

   a. Find all generalizations of a pattern.

      i. Replacing values with variables.

        To generalize a pattern we need to replace a set of values (in particular a singleton) that occur in the pattern's condition with a variable. We assume that the cost of the replacement of a set of values is 1.

        Let PS be the set of all patterns, p the pattern to be generated, $cond(p)$ the set of all conditions of p, and $comp(c)$ the set of components (constants, a set of constants or variables) of a pattern condition c to be modified, i.e. generalized or specialized. The cost of the generalization of a pattern is

$$T(n) = \sum_{c \in cond(p)} \sum_{l \in comp(c)} 1 = |cond(p)||comp(c)| \leq C_1 C_2 = O(1),$$

        where $C_1 = \max\{|cond(p)| : p \in PS\}$, $C_2 = max\{|comp(c)| : c \in cond(p), p \in PS\}$. Values C_1 and C_2 are small and do not depend on the data size.

     ii. Removing conditions.

        To generalize a pattern we need to scan all of the pattern's conditions in order to remove one of them. We assume that the cost of the removal of a condition is 1.

        Let p be a pattern to be specialized. The cost of the generalization of a pattern is

$$T(n) = \sum_{c \in cond(p)} 1 = |cond(p)| \leq C = O(1),$$

        where $C = \max\{|cond(p)| : p \in PS\}$ is small and does not depend on the data size.

   b. Find all specializations of a pattern.

      i. Replacing variable with set values.

        To specialize a pattern we need to replace its condition variable with a list of values. We assume that the cost of the replacement of a variable is 1.

        Let $val(V)$ be the values set of a variable V, and S_V the family of sets of values taken into account during the replacement of a variable V. The cost of the specialization of a pattern is

$$T(n) = \sum_{c \in cond(p)} \sum_{V \in comp(c)} \sum_{l \in S_V} 1 \leq \sum_{c \in cond(p)} \sum_{V \in comp(c)} (2^{|val(V)|} - 2) =$$

$$|cond(p)||comp(c)|(2^{|val(V)|} - 2) \leq C_1 C_2 (2^{C_3} - 2) = O(1),$$

where $C_1 = \max\{|cond(p)| : p \in PS\}$, $C_2 = \max\{|comp(c)| : c \in cond(p), p \in PS\}$, $C_3 = \max\{|val(V)| : V \in var(c), c \in cond(p), p \in PS\}$. In a pessimistic case, we have $S_V = P(val(V)) \setminus \{\emptyset, val(V)\}$.[6] C_1 and C_2 are small and do not depend on the data size, and neither does C_3 since we assume that the data is discretized.

  ii. Adding conditions.

To specialize a pattern we need to choose a new condition. We assume that the cost of the choice of a condition is 1.

Let $cond_i(p)$ be the set of all conditions to be generated for a pattern p at a given level i. The cost of the specialization of a pattern is

$$T(n) = \sum_{c \in cond_i(p)} 1 = |cond_i(p)| \leq C = O(1),$$

where $C = \max\{|cond_i(p)| : p \in PS\}$ is small and does not depend on the data size.

3. Check if the target objects satisfy a pattern.

If a pattern is only constructed by adding or removing conditions, it is enough to scan $rlt_{gen}(o)$ to check if o satisfies the pattern. If any condition of a pattern is generalized or specialized, we need to additionally scan $rlt(o)$ to check if o satisfies the condition. However, we assume that background objects from $rlt(o)$ are associated with the corresponding objects from $rlt_{gen}(o)$. Thanks to this, there is no need to scan the whole $rlt(o)$. Hence, we can ignore the cost of finding a background object to satisfy a given condition and we assume that the cost of the verification of a condition is 1.

Let $O \subseteq U_{D_T}$ be a set of objects for which a pattern is to be checked, and $mod(c)$ the set of all conditions derived from a condition c by applying the *special* or *general* function. The cost of checking the pattern satisfaction is

$$T(n) = \sum_{o \in O} \sum_{o' \in rlt_{gen}(o)} \sum_{c \in cond(p)} \sum_{c' \in mod(c)} 1 =$$

$$|o \in O||rlt_{gen}(o)||\{c \in cond(p)\}||mod(c)| \leq nC_1 C_2 C_3 = O(n),$$

---

[6]$P(X)$ is the power set of X.

where $C_1 = \max\{|rlt_{gen}(o)| : o \in U_{D_T}\}$, $C_2 = |cond(p)|$, $C_3 = \max\{|mod(c)| : c \in cond(p)\}$. When the whole pattern is only constructed by adding or removing conditions, then $C_3 = 1$.

The operations from points 1–3 are independent, thus the complexity of the generation of a pattern is $O(1) + O(1) + O(n) = O(n)$. We assumed that the database is representative, hence we obtain that the number of patterns does not depend on the data size. Thus, the complexity of the generation of a pattern set PS is

$$|PS|O(n) = O(n),$$

where $FP'_p = special(p, m)$ or $FP'_p = general(p, m)$.

Based on the above analysis one can immediately show the pattern generation approach's scalability with respect to the data size.

Definition 4.1 (Algorithm's scalability) An algorithm is scalable with respect to data size n if it has a linear time complexity.

Proposition 4.1 *An algorithm for pattern generation based on the framework is scalable.*

4.5   Conclusions

This chapter has introduced specialized frameworks indented for association discovery and classification rules mining from relational data. They both are based on the general framework for mining relational data. The structure for storing relational data in this framework is an information system that is constructed by adapting the notion of standard information system. Information granules derived from the information system are used to construct relational patterns such as frequent patterns, association rules, and classification rules. The frameworks enable to define new algorithms as well as redefine existing ones for generating relational patterns. A granular representation of relational data can be a platform for not only different rule mining algorithms but also for different tasks, e.g. association discovery and classification rules mining.

Chapter 5
Rough-Granular Computing

5.1 Introduction

Rough set theory [71] as a useful tool to deal with imprecise data is often considered as one of basic techniques of granular computing [9, 72]. In this view, granules are formed by means of rough inclusions as classes of objects close to a specified center of the granule to a given degree. Formally, they resemble neighborhoods formed with respect to a certain metric.

In recent years, one can observe a trend in data mining towards the application of granular computing based on the rough set approach. This newly emerging approach is called rough-granular computing [85, 89].

Techniques of granular computing, especially rough sets, have widely been applied in the field of data mining (see, e.g. [9, 72, 85]). Methods of rough sets have also found application in mining data stored in multiple tables, i.e. relational data mining. Namely, it has found application in tasks such as eliminating unimportant data (see, e.g. [88]); the analysis of invalid, missing, and indistinguishable data (see, e.g. [60, 66]); reducing data size (see, e.g. [65]); relational classification rules generation (see, e.g. [65, 67, 90]).

This chapter develops a rough-granular computing framework for mining relational data [37]. To this end, the tolerance rough set model [82, 88] is adapted. Two ways for constructing the universe from relational data are introduced: the universe constructed from granules directly derived from relational data and the one constructed from information granules being a generalized representation of relational data. The framework combines advantages of both granular computing and rough sets. Due to applying granular computing methods, one can overcome the problems of relational data representation and the search space limitation. The application of rough sets makes it possible to deal with uncertainty in relational data.

The remaining of the chapter is organized as follows. Section 5.2 provides a rough-granular computing model developed for single table data. Sections 5.3 and 5.4 introduce approximation spaces for relational granules and generalized relational granules, respectively. Section 5.5 provides conclusion remarks.

© Springer International Publishing AG 2017

P. Hońko, *Granular-Relational Data Mining*, Studies in Computational
Intelligence 702, DOI 10.1007/978-3-319-52751-2_5

5.2 Rough-Granular Computing for Single Table Data

Rough-granular computing can be viewed as rough set theory interpreted in the framework of granular computing and applied to discovering knowledge from databases. Elementary granules in this approach are represented by indiscernibility or similarity classes. Higher level granules, which correspond to rough approximations, are constructed based on elementary granules that totally (lower approximation) or partially (upper approximation) belong to the concept under consideration. These granules are the basis for discovering relevant patterns (e.g. classification rules) describing the concept.

Depending on the data type and the task to be performed, different rough set models can be used as the core of rough-granular computing. In this work, the tolerance rough set model [82, 88] is taken due to its flexibility in tuning parameters.

Definition 5.1 [82] (*Approximation space*) A parameterized approximation space $AS_{\#,\$}$ for an information system $IS = (U, A)$ is defined by $AS_{\#,\$} = (U, I_\#, \nu_\$)$, where

- U is a non-empty set of objects,
- $I_\# : U \to P(U)$ is an uncertainty function,
- $\nu_\$: P(U) \times P(U) \to [0, 1]$ is a rough inclusion function.

For every object, the uncertainty function defines a set of similarly described objects (elementary granule). The function can be defined as follows.

Definition 5.2 (cf. [82]) (*Uncertainty function*) Let $IS = (U, A)$ be an information system. An uncertainty function $I_{B,\varepsilon}$ is defined by

$$I_{B,\varepsilon}(x) = \bigcap_{a \in B} I_{a,\varepsilon_a}(x)$$

where $x \in U$, $B \subseteq A$, $\varepsilon = (\varepsilon_a : a \in B)$ is a vector of thresholds such that $\varepsilon_a \geq 0$ for $a \in B$, $I_{a,\varepsilon_a}(x) = \{y \in U : d_a(x, y) \leq \varepsilon_a\}$, and $d_a : U \times U \to [0, \infty)$ is a distance measure.

The rough inclusion function defines the degree of inclusion of a set X in a set Y, where $X, Y \subseteq U$. Depending on its definition, the rough inclusion function can satisfy different properties. The following properties will be considered.

1. $\underset{A,B \subseteq U}{\forall}\ A \subseteq B \Rightarrow \nu_\$(A, B) = 1\ (p_1)$,
2. $\underset{A,B \subseteq U}{\forall}\ \nu_\$(A, B) = 1 \Leftrightarrow A \subseteq B\ (p_2)$,
3. $\underset{A,B,C \subseteq U}{\forall}\ \nu_\$(B, C) = 1 \Rightarrow \nu_\$(A, B) \leq \nu_\$(A, C)\ (p_3)$,
4. $\underset{A,B,C \subseteq U}{\forall}\ B \subseteq C \Rightarrow \nu_\$(A, B) \leq \nu_\$(A, C)\ (p_4)$,
5. $\underset{A,B \subseteq U}{\forall}\ \nu_\$(A, B) = 0 \Leftrightarrow A \cap B = \emptyset\ (p_5)$.

We call $\nu_\$$ rough inclusion function (RIF), quasi-rough inclusion function (q-RIF), or weak quasi-rough inclusion function (weak q-RIF) if it satisfies properties p_2 and p_3, p_1 and p_3, or p_1 and p_4, respectively [31]. Property p_5 is optional (see Sect. 9.2.2).

The following rough inclusion functions will be used.

Definition 5.3 [82] (*Rough inclusion functions*) The rough inclusion $\nu_{l,u}(X, Y)$ of a set X in a set Y is defined by

$$\nu_{l,u}(X, Y) = f_{l,u}(\nu_{SRI}(X, Y)), \text{ where } f_{l,u}(t) = \begin{cases} 0 & \text{if } 0 \leq t \leq l \\ \frac{t-l}{u-l} & \text{if } l < t < u \\ 1 & \text{if } t \geq u \end{cases},$$

$$0 \leq l < u \leq 1 \text{ and } \nu_{SRI}(X, Y) = \begin{cases} \frac{card(X \cap Y)}{card(X)} & \text{if } X \neq \emptyset \\ 1 & \text{if } X = \emptyset \end{cases} \text{ is the standard rough inclusion.}$$

Note that if $l = 0$ and $u = 1$, then the rough inclusion $\nu_{l,u}$ is equivalent to the standard rough inclusion ν_{SRI}.

The lower and upper approximations (higher level granules) of a concept are defined as follows.

Definition 5.4 [82] (*Approximations of a subset in $AS_{\#,\$}$*) For an approximation space $AS_{\#,\$} = (U, I_\#, \nu_\$)$ and any subset $X \subseteq U$, the lower and the upper approximations are defined respectively by

$$LOW(AS_{\#,\$}, X) = \{x \in U : \nu_\$(I_\#(x), X) = 1\},$$

$$UPP(AS_{\#,\$}, X) = \{x \in U : \nu_\$(I_\#(x), X) > 0\}.$$

Symbols #, \$ denote vectors of parameters which can be tuned in the process of concept approximation.

Definition 5.5 (cf. [71]) (*Rough set*) Let $AS_{\#,\$} = (U, I_\#, \nu_\$)$ be an approximation space. The tolerance rough set of a subset $X \subseteq U$ is defined by the pair $(LOW(AS_{\#,\$}, X), UPP(AS_{\#,\$}, X))$.

Example 5.1 Consider the database from Example 2.1. Let $AS_{(B,\varepsilon),(l,u)} = (U, I_{B,\varepsilon}, \nu_{l,u})$ be an approximation space, where $U = \{o_i : 1 \leq i \leq 7\}$, o_i correspond to *i*th object from table *customer*, $B = \{age, income\}$, $\varepsilon = (\varepsilon_{age}, \varepsilon_{income}) = (5, 500)$, the distance measure is $d(x, y) = |a(x) - a(y)|$, $l = 0.33$, $u = 0.67$.

Let $X_1 = \{1, 2, 4, 5, 6\}$ be the set (i,e. concept) to be approximated.

The table below shows the similarity classes and their rough inclusion degrees in X.

We obtain the following approximations (higher level granules) LOW $(AS_{(B,\varepsilon),(l,u)}, X) = \{2, 5, 6\}$, $UPP(AS_{(B,\varepsilon),(l,u)}, X) = \{1, 2, 5, 6, 7\}$.

$o_i \in U$	$I_{B,\varepsilon}(o_i)$	$\nu_{l,u}(I_{B,\varepsilon}(o_i), X)$
1	$\{1, 7\}$	0.5
2	$\{2, 6\}$	1
3	$\{3, 4, 7\}$	0.33
4	$\{3, 4, 7\}$	0.33
5	$\{5, 6\}$	1
6	$\{5, 6\}$	1
7	$\{1, 3, 4, 7\}$	0.5

5.3 Approximation Space for Relational Granules

This section introduces an approximation space for granules directly derived from relational data.

Consider a finite database $D = T \cup B$, where $T = U_{D_T}$ and $B = U_{D_B}$ are, respectively, the sets of target and background relations of D. Let $IS_D = (U_D, A_D)$ be an information system of a given database D, where $U_D = U_{D_T} \cup U_{D_B}$.

Definition 5.6 (*Approximation space $AS_{\#,\$}^i$*) An approximation space $AS_{\#,\$}^i$ for a database $D = T \cup B$ represented by the information system $IS_D = (U_D, A_D)$ is defined by $AS_{\#,\$}^i = (U^i, I_\#, \nu_\$)$, where

- $U^i = \{(o, rlt^i(o)) : o \in U_T\}$ is a non-empty set of granules,
- $I_\# : U^i \to P(U^i)$ is an uncertainty function,
- $\nu_\$: P(U^i) \times P(U^i) \to [0, 1]$ is a rough inclusion function.

The component of an approximation space that needs to be adapted to operating on granules is the uncertainty function. It can be constructed based on similarity measures.

Typical measure can be used for attribute values.

Definition 5.7 (*Similarity of values*) Let $v, v' \in V_a$ be values of an attribute. The similarity of values v and v' is calculated as

$$sim_a(v, v') = \begin{cases} (v = v') & \text{if } a \text{ is nominal,} \\ \frac{|v - v'|}{|max V_a - min V_a|} & \text{if } a \text{ is numerical,} \end{cases}$$

where $(v = v')$ returns 1 if $v = v'$ and 0 otherwise.

The first step is to measure the similarity of objects of the same relation.

Definition 5.8 (*Similarity of objects*) Let o and o' be relational objects constructed over a relation R. The similarity of objects o and o' for attribute subset $B \subseteq R.A$ is computed as follows

$$sim_B(o, o') = \begin{cases} \frac{\sum_{a \in B} sim_a(a(o), a(o'))}{|B|} & \text{if } B \neq \emptyset, \\ 0 & \text{if } B = \emptyset, \end{cases}$$

Since a target object may be joined with more than one object of the same relation, the measure that operates on sets of objects is introduced.

Definition 5.9 (*Similarity of sets*) Let S_R and S'_R be sets of relational objects constructed over a relation R such that $|S_R| \leq |S'_R|$. The similarity of sets S_R and S'_R is computed as follows

$$R\_sim_B(S_R, S'_R) = \frac{max \left\{ \sum_{i=1}^{|S_R|} sim_B(P[i], P'[i]) : P' \in perm(S'_R) \right\}}{|S'_R|}$$

where $perm(S)$ is the set of all permutations of a set S, and P is a certain permutation of S_R.

Due to operating on permutations the measure is suitable for relatively small sets. The next measure operate on sets including objects of different relations.

Let $\mathcal{B} = \{B_R : R \in S \cap S', B \subseteq R.A\}$ where S and S' are sets of relational objects.

Definition 5.10 (*Similarity of sets of relational objects*) The similarity of sets S and S' of relational objects is computed as follows

$$S\_sim_{\mathcal{B}}(S, S') = \frac{\sum_{R \in rel(S) \cap rel(S')} R\_sim_{B_R}(S_R, S'_R)}{|rel(S) \cup rel(S')|}$$

Finally, one can define a measure operating on granules.

Definition 5.11 (*Similarity of granules*) Let $g = (o, rlt(o))$ and $g' = (o', rlt(o'))$ be granules such that $o, o' \in U_{D_T}$. The similarity of granules g and g' is computed as follows

$$g\_sim_{\mathcal{B}}(g, g') = S\_sim_{\mathcal{B}}(\{o\} \cup rlt(o), \{o'\} \cup rlt(o')).$$

The above measure does not take depth levels into account. They are essential if the same relation can be found at different levels. This situation takes place when related sets are constructed in a recursive way. The measure defined below treats each level separately.

Let g_j be the granule $g = (o, rlt^i(o))$ limited to depth level $j \leq i$. For $j = 0$ we assume that $g_0 = o$. Let also $\mathcal{B}_j$ be $\mathcal{B}$ limited to depth level $j \leq i$.

Definition 5.12 (*Similarity of granules w.r.t. depth levels*) Let $g = (o, rlt^i(o))$ and $g' = (o', rlt^i(o'))$ be granules such that $o, o' \in U_{D_T}$. The similarity of granules g and g' with respect to depth levels is computed as follows

$$g^i\_sim_{\mathcal{B}}(g, g') = \frac{\sum_{j=0}^{i} S\_sim_{\mathcal{B}_j}(g_j, g'_j)}{i+1}.$$

Example 5.2 Consider the following granules constructed based on the database from Example 2.1: $g_1 = (o_1, rlt^3(o_1)) = (c_1, \{m_1, p_1, p_2, c_5, p'_1, p'_3\}), g_2 = (o_4, rlt^3(o_4)) = (c_4, \{, m_2, p_5, p_6, c_6, p'_2, p'_6, p_7\}).$[1]

Let $\mathscr{B} = \{B_c, B_m, B_p, B_{p'}\}$ where $B_c = \{age, income\}, B_m = \emptyset$ (relation *married_to* does not include descriptive attributes), $B_p = \{prod\_id, amount\}$ and $B_{p'} = \{name\}$.

We obtain $g\_sim_{\mathscr{B}}(g_1, g_2) = \frac{R\_sim_{B_c} + R\_sim_{B_m} + R\_sim_{B_p} + R\_sim_{B_{p'}}}{4} = \frac{0.64+1+0.25+0}{4} = 0.47.$

Taking the depth levels into account, we obtain $g^i\_sim_{\mathscr{B}}(g_1, g_2) = \frac{S\_sim_{\{B_c\}}(\{c_1\},\{c_4\})}{4}$

$+ \frac{S\_sim_{\{B_m, B_p\}}(\{m_1, p_1, p_2\}, \{m_3, p_5, p_6\}) + S\_sim_{\{B_c, B_{p'}\}}(\{c_5, p'_1, p'_3\}, \{c_6, p'_2, p'_6\}) + S\_sim_{\{B_p\}}(\emptyset, \{p_7\})}{4} =$

$\frac{0.6+0.69+0.34+0}{4} = 0.41.$

Having a similarity measure that operates on granules, one can define the uncertainty function.

Definition 5.13 (*Uncertainty function in* $AS^i_{\#,\$}$) The uncertainty function $I_{\mathscr{B},\varepsilon}$ in an approximation space $AS^i_{\#,\$}$ is defined as follows

$$I_{\mathscr{B},\varepsilon}(g) = \{g' \in U^i : g\_sim_{\mathscr{B}}(g, g') \geq \varepsilon\},$$

where $\varepsilon \in (0, 1]$ is a similarity threshold.

In the above definition, $g\_sim_{\mathscr{B}}$ can be replaced with $g^i\_sim_{\mathscr{B}}$ to take into account depth levels. The similarity can also be verified at each depth level separately as it is done in the following function.

Definition 5.14 (*Uncertainty function operating on depth levels in* $AS^i_{\#,\$}$) Let $\varepsilon = \{\varepsilon_j : j = 1, \ldots, i\}$, where $\varepsilon_j \in (0, 1]$ is a similarity threshold for depth level j. The uncertainty function $I^i_{\mathscr{B},\varepsilon}$ in an approximation space $AS^i_{\#,\$}$ that operates on depth levels is defined as follows.

$$I^i_{\mathscr{B},\varepsilon} = \bigcap_{j=0}^{i} I_{\mathscr{B},\varepsilon_j}(g_j).$$

Example 5.3 Consider the following approximation spaces for the database (without relation *married_to*) from Example 2.1: $AS^0_{(B_0,\varepsilon_0),(l,u)} = (U^0, I_{B_0,\varepsilon_0}, \nu_{l,u})$ and $AS^1_{(\mathscr{B},),(l,u)} = (U^1, I_{\mathscr{B},\varepsilon}, \nu_{l,u})$, where $l = 0.33, u = 0.67, B_0 = \{age, income\}$, $\varepsilon_0 = 0.6, \mathscr{B} = \{B_0, B_1\}, = \{\varepsilon_0, \varepsilon_1\}, B_1 = \{prod\_id, amount\}, \varepsilon_1 = 0.25.$[2]
Universe U^1 consists of the following granules $g_1 = (c_1), \{p_1, p_2\}), g_2 = (c_2),$ $\{p_3, p_4\}), g_3 = (c_3), \{p_8\}), g_4 = (c_4), \{p_5, p_6\}), g_5 = (c_5, \emptyset), g_6 = (c_6, \{p_7\}), g_7 = (c_7, \emptyset).$

---

[1]Symbols c_i, m_i, p_i, p'_i denote the i-th object of tables *customer, married_to, purchase, product*, respectively.

[2]Here, attribute *prod_id* is treated as nominal.

Let $X_1 = \{1, 2, 4, 5, 6\}$ be the set to be approximated.

The table below shows the similarity classes and their rough inclusion degrees in X.

$g \in U^0$	$I_{B_0,\varepsilon_0}(g)$	$\nu_{l,u}(I_{B_0,\varepsilon_0}(x), X)$	$g \in U^1$	$I_{\mathscr{B},\varepsilon}(g)$	$\nu_{l,u}(I_{\mathscr{B},\varepsilon}(g), X)$
1	$\{1, 3, 4\}$	0.67	1	$\{1, 3, 4\}$	0.67
2	$\{2, 3, 4, 5, 6, 7\}$	0.67	2	$\{2, 3, 4\}$	0.67
3	$\{1, 2, 3, 4, 7\}$	0.6	3	$\{1, 2, 3, 4\}$	0.75
4	$\{1, 2, 3, 4, 7\}$	0.6	4	$\{1, 2, 3, 4\}$	0.75
5	$\{2, 5, 6\}$	1	5	$\{5\}$	1
6	$\{2, 5, 6\}$	1	6	$\{6\}$	1
7	$\{2, 3, 4, 7\}$	0.5	7	$\{7\}$	0

We obtain the following approximations $LOW(AS^0_{(B_0,\varepsilon_0),(l,u)}, X) = \{1, 2, 5, 6\}$, $UPP(AS^0_{(B_0,\varepsilon_0),(l,u)}, X) = U^0, LOW(AS^1_{(\mathscr{B},),(l,u)}, X) = \{1, 2, 3, 4, 5, 6\}$, $UPP(AS^1_{(\mathscr{B},),(l,u)}, X) = \{1, 2, 3, 4, 5, 6\}$.

5.4 Approximation Space for Generalized Relational Granules

This section introduces an approximation space for information granules being a generalized representation of relational data.

The universe defined as in the previous subsection makes it possible to apply a wide range of uncertainty functions. Furthermore, the size of granules from the universe can be adjusted by changing the depth level.

In order to additionally limit the size of the universe, granules are constructed based on generalized descriptions of target objects. To obtain a general (i.e. abstract) description of a target object itself and its related set, they both are generalized.

Definition 5.15 (*Approximation space $genAS^i_{\#,\$}$*) An approximation space $genAS^i_{\#,\$}$ for a database $D = T \cup B$ represented by the information system $IS_D = (U_D, A_D)$ is defined by $genAS^i_{\#,\$} = (U^i_{gen}, I_\#, \nu_\$)$, where

- $U^i_{gen} = \{(o_{gen}, rlt^i_{gen}(o)) : o \in U_T\}$ is a non-empty set of granules,
- $I_\# : U^i_{gen} \to P(U^i_{gen})$ is an uncertainty function,
- $\nu_\$: P(U^i_{gen}) \times P(U^i_{gen}) \to [0, 1]$ is a rough inclusion function.

To compute the similarity of generalized objects of the same relation, the measures from Definitions 5.7 and 5.8 can be used. Attributes that are replaced in a generalized object with variables are treated as nominal.

The mentioned measures can be used if a syntactic comparison of generalized objects in sufficient. Otherwise the following measure can be applied.

Definition 5.16 (*Semantic similarity of objects*) Let o_{gen} and o'_{gen} be relational objects constructed over the same relation. The semantic similarity of objects o_{gen} and o'_{gen} for attribute subset B is computed as follows

$$sim'_B(o_{gen}, o'_{gen}) = \begin{cases} 1 & \text{if} \quad \exists_\sigma o_{gen}\sigma = o'_{gen} \wedge o'_{gen}\sigma^{-1} = o_{gen}; \\ 0 & \text{otherwise.} \end{cases}$$

Example 5.4 Let $S = rlt^1_{gen}(o_4) = \{purchase(B, A, C, 1, \_), purchase(B', A, C', 3, \_)\}$ and $S' = rlt^1_{gen}(o_6) = \{purchase(B, A, C, 3, \_)\}$, where o_4 and o_6 correspond to customers 4 and 5 from Example 2.1, respectively (relation *married_to* not taken). Let $B_0 = \{purchase.id, cust\_id, prod\_id, amount\}$.

We obtain $sim_{B_0}(purchase(B, A, C, 1, \_), purchase(B, A, C, 3, \_)) = 0.75$ and $sim_{B_0}(purchase(B', A, C', 3, \_), purchase(B, A, C, 3, \_)) = 0.5$, hence $R\_sim_B(S, S') = 0.38$. Using measure sim'_{B_0} we obtain $sim'_{B_0}(purchase(B', A, C', 3, \_), purchase(B, A, C, 3, \_)) = 1$ ($\sigma = \{B'/B, C'/C\}$), hence $R\_sim'_{B_0}(S, S') = 0.5$.

Definition 5.17 (*Extended semantic similarity of objects*) Let o_{gen} and o'_{gen} be relational objects constructed over the same relation. The extended semantic similarity of objects o_{gen} and o'_{gen} for attribute subset B is computed as follows

$$sim''_B(o_{gen}, o'_{gen}) = \begin{cases} 1 & \text{if} \quad \exists_\sigma o_{gen}\sigma = o'_{gen} \wedge o'_{gen}\sigma^{-1} = o_{gen}; \\ 0.5 & \text{if} \quad \exists_\sigma o_{gen}\sigma = o'_{gen} \vee o'_{gen}\sigma = o_{gen}; \\ 0 & \text{otherwise.} \end{cases}$$

Example 5.5 For illustrative purposes, consider a relation *filmmakers(scenarist, director, producer)* and its three relational objects $o = filmmakers(1, 2, 3), o' = filmmakers(2, 2, 3), o'' = filmmakers(4, 5, 5)$. Let the generalized objects be the following $o_{gen} = filmmakers(A, B, C), o'_{gen} = filmmakers(A, A, B), o''_{gen} = filmmakers(A, B, B)$. Let $B_0 = filmmakers.A$.
Using measure sim'_{B_0} we obtain that the similarities of any pair of the objects is 0. For measure sim''_{B_0} we have $sim''_{B_0}(o_{gen}, o'_{gen}) = 0.5$ ($\sigma = \{B/A, C/B\}$, $o_{gen}\sigma = o'_{gen}$ and $o'_{gen}\sigma^{-1} \neq o_{gen}$), $sim''_{B_0}(o_{gen}, o''_{gen}) = 0.5$ ($\sigma' = \{C/B\}$, $o_{gen}\sigma' = o''_{gen}$ and $o''_{gen}\sigma'^{-1} \neq o_{gen}$), and $sim''_{B_0}(o'_{gen}, o''_{gen}) = 0$.

If during generalization only variables introduced at higher levels are taken into account, then one may obtain a generalized related set with repetitions.

Definition 5.18 (*Generalized related set with repetitions*) A generalized related set with repetitions is defined as follows $rrlt_{gen}(o) = \{(o_1, l_1), \ldots, (o_n, l_n) : rlt_{gen}(o) = \{o_1, \ldots, o_n\}, n \leq |rlt_{gen}(o)|\}$ where l_j is the number of repetitions of object o_j.

Let $minmax(v, v') = \frac{min\{v,v'\}}{max\{v,v'\}}$ where v and v' are positive numbers.

Definition 5.19 (*Similarity of objects with repetitions*) The similarity of objects with repetitions (o, l) and (o', l') is computed as follows

$$sim((o, l), (o', l')) = minmax(l, l')sim(o, o'),$$

where $sim(o, o')$ is any similarity measure operating on o and o'.

During generalization a value that occurs in a object can be replaced with a set of values. To handle such sets the following measures are introduced.

Definition 5.20 (*Similarity of sets*) Let V and V' be subsets of V_a where a is an attribute. The similarity of sets V and V' is computed as follows

$$sim(V, V') = \begin{cases} \frac{|V \cap V'|}{|V \cup V'|} & \text{if } a \text{ is nominal;} \\ minmax(avgV, avgV') & \text{if } a \text{ is numerical,} \end{cases}$$

where $avgV$ is the average of values from V.[3]

To take into account the similarity of sets in terms of size, the following measure is used.

Definition 5.21 (*Similarity of sets w.r.t. set size*) Let V and V' be subsets of V_a where a is an attribute. The similarity of sets V and V' with respect to the set size is computed as follows

$$sim'(V, V') = \big(sim(V, V') + minmax(|V|, |V'|)\big)/2.$$

Example 5.6 Let $rrlt^1_{gen}(o_2) = \{(purchase(B, A, C, 1, \_), 2)\}$ and $rrlt^1_{gen}(o_3) = \{(purchase(B, A, C, 1, \_), 1)\}$, where o_2 and o_3 correspond to customers 2 and 3 from Example 2.1, respectively. Let $B_0 = \{purchase.id, cust\_id, prod\_id, amount\}$.

We obtain $sim_{B_0}((purchase(B, A, C, 1, \_), 2), (purchase(B, A, C, 1, \_), 1)) = 1/2$. Consider also different generalizations $rlt^1_{gen}(o_2) = \{purchase(B, A, \_, \{1, 3\}, \_))\}$ and $rlt^1_{gen}(o_3) = \{purchase(B, A, \_, 4, \_))\}$. Let $B_0 = \{prod\_id\}$. Using the measure from Definition 5.20 we obtain $sim_{B_0}(purchase(B, A, C, \{1, 3\}, \_), purchase(B, A, C, 4, \_)) = sim(\{1, 3\}, \{4\}) = 0$. For the measure from Definition 5.21 we have $sim_{B_0}(purchase(B, A, C, \{1, 3\}, \_), purchase(B, A, C, 4, \_)) = sim'(\{1, 3\}, \{4\}) = 0.25$. Customer 2 bought two products, whereas customer 3—one. It means that their purchases in terms of size are similar to degree 0.5.

Example 5.7 Consider the following approximation spaces for the database (without relation *married_to*) from Example 2.1: $AS^0_{(B_0,),(l,u)} = (U^0_{gen}, I_{B_0,0}, \nu_{l,u})$ and $AS^1_{(\mathcal{B},),(l,u)} = (U^1_{gen}, I_{\mathcal{B},\varepsilon}, \nu_{l,u})$, where $l = 0.33, u = 0.67, B_0 = \{age, income\}$, $\varepsilon_0 = 0.6, \mathcal{B} = \{B_0, B_1\}, = \{\varepsilon_0, \varepsilon_1\}, B_1 = \{prod\_id, amount\}, \varepsilon_1 = 0.25$.

---

[3]The measure can be used for sets of positive numbers only.

Let attributes *age* and *income* be generalized as follows $age_1 = \{25\text{-}30\}$, $age_2 = \{30\text{-}35\}$, $age_3 = \{36\text{-}40\}$, $inc_1 = \{1500\text{-}2000\}$, $inc_2 = \{2500\text{-}3000\}$. Let relation *purchase* be generalized using the aggregation approach with respect to attribute *prod_id*. Universe U_{gen}^1 consists of the following elements:

$g_1 = (customer(A, \_, age_3, \_, inc_1, yes), purchase(B, A, \_, \{1, 3\}, \_))$,
$g_2 = (customer(A, \_, age_2, \_, inc_2, yes), purchase(B, A, \_, \{1, 3\}, \_))$,
$g_3 = (customer(A, \_, age_2, \_, inc_1, no), purchase(B, A, \_, 4, \_))$,
$g_4 = (customer(A, \_, age_2, \_, inc_1, yes), purchase(B, A, \_, \{2, 6\}, \_))$,
$g_5 = (customer(A, \_, age_1, \_, inc_2, yes), \emptyset)$,
$g_6 = (customer(A, \_, age_1, \_, inc_1, yes), purchase(B, A, \_, 5, \_))$,
$g_7 = (customer(A, \_, age_2, \_, inc_1, no), \emptyset)$.

Let $X_1 = \{1, 2, 4, 5, 6\}$ be the set to be approximated.

The table below shows the similarity classes and their rough inclusion degrees in X.

$g \in U^0$	$I_{B_0,\varepsilon_0}(g)$	$\nu_{l,u}(I_{B_0,\varepsilon_0}(x), X)$	$g \in U^1$	$I_{\mathscr{B},\varepsilon}(g)$	$\nu_{l,u}(I_{\mathscr{B},\varepsilon}(g), X)$
1	$\{1, 3, 4, 6, 7\}$	0.6	1	$\{1, 3, 4\}$	0.67
2	$\{2, 3, 4, 5, 6\}$	0.8	2	$\{2, 3, 4, 6\}$	0.75
3	$\{1, 2, 3, 4, 6, 7\}$	0.67	3	$\{1, 2, 3, 4, 6\}$	0.8
4	$\{1, 2, 3, 4, 6, 7\}$	0.67	4	$\{1, 2, 3, 4, 6\}$	0.8
5	$\{2, 5, 6\}$	1	5	$\{5\}$	1
6	$\{1, 2, 3, 4, 5, 6, 7\}$	0.71	6	$\{1, 2, 3, 4, 6\}$	0.8
7	$\{1, 3, 4, 6, 7\}$	0.6	7	$\{7\}$	0

We obtain the following approximations $LOW(AS_{(B_0,\varepsilon_0),(l,u)}^0, X) = \{2, 3, 4, 5, 6\}$, $UPP(AS_{(B_0,\varepsilon_0),(l,u)}^0, X) = U^0$, $LOW(AS_{(\mathscr{B},),(l,u)}^1, X) = \{1, 2, 3, 4, 5, 6\}$, $UPP(AS_{(\mathscr{B},),(l,u)}^1, X) = \{1, 2, 3, 4, 5, 6\}$.

5.5   Conclusions

This chapter has introduced a rough-granular computing framework for mining relational data. Two ways for constructing the universe based on relational data have been defined: the universe constructed from granules directly derived from relational data and the one constructed from information granules being a generalized representation of relational data.

The rough-granular computing framework enables to describe in an approximate way concepts derived from relational data. The framework can also be used as a tool in the process of discovering patterns from relational data. Namely, it can be embedded in the classification rules mining framework (Chap. 4) to improve classification of uncertain relational data.

Part II
Description Language Based Approach

Chapter 6
Compound Information Systems

6.1 Introduction

The goal of this chapter is to provide a general framework for analyzing and processing relational data in a granular computing environment [41]. This work can be treated as an extension of a granular computing based framework intended to handle propositional data, i.e. data stored in a single table [83]. Two information systems for storing relational data are introduced. They both combine universes of information systems (each corresponding to one database table) into one universe. The first system, called compound information system, allows all combinations of objects from the particular universes, whereas the second one, called constrained compound information system, reflects relationships that exist in the database. The chapter also extends an attribute-value language to express relationships among objects as well as a language for granule description to express information granules derived from relational data, called relational information granules.

Relational information granules are the basis for constructing patterns to be discovered from relational data. Thanks to this approach, the patterns can be formed over a simpler language (i.e. an extended attribute-value language) compared with a relational one, but they are able to preserve expressiveness of their relational counterparts.

The remaining of the chapter is organized as follows. Section 6.2 introduces relational information granules. Sections 6.3 and 6.4 develop a compound information system and constrained compound information system, respectively. Section 6.5 investigates the consistency and completeness of the approach. Section 6.6 provides concluding remarks.

© Springer International Publishing AG 2017

P. Hońko, *Granular-Relational Data Mining*, Studies in Computational
Intelligence 702, DOI 10.1007/978-3-319-52751-2_6

6.2   Information Granules

This section introduces a granular computing based framework which is constructed based on definitions from [83].

In the traditional database theory, attribute values are the primitives for data mining. In granular computing theory, in turn, the primitives are defined by granules of entities. Data can be transformed into granules by applying a partition or covering of the universe. In the former case, granules are defined by equivalence classes, and in the latter case—by similarity classes.

This chapter uses a framework in which granules are constructed by applying a partition of the universe.[1] More precisely, each attribute value forms one granule consisting of objects that share the value.

An information granule is represented by an expression of the form (*name*, *content*), where *name* is the granule identifier and *content* is a set of objects identified by *name* [89]. To construct granules, logical formulas over some language are used, i.e. granule description language. Namely, granules are defined by formulas which are used to express the properties of the objects from the granules.

It is assumed that an information system $IS = (U, A)$ is given along with the following:

- a set of formulas Φ over some language.
- a function $SEM : \Phi \rightarrow P(U)$.

Definition 6.1 (*Syntax and semantics of L_{IS}*) The syntax and semantics of the language L_{IS} are defined recursively by[2]

1. $a \in A, v \in V_a \Rightarrow (a, v) \in L_{IS}$ (an atomic formula) and $SEM_{IS}(a, v) = \{x \in U : a(x) = v\}$[3];
2. $\alpha \in L_{IS} \Rightarrow \neg \alpha \in L_{IS}$ and $SEM_{IS}(\neg \alpha) = U \backslash SEM_{IS}(\alpha)$;
3. $\alpha_1, \alpha_2 \in L_{IS} \Rightarrow \alpha_1 \wedge \alpha_2 \in L_{IS}$ and $SEM_{IS}(\alpha_1 \wedge \alpha_2) = SEM_{IS}(\alpha_1) \cap SEM_{IS}(\alpha_2)$;
4. $\alpha_1, \alpha_2 \in L_{IS} \Rightarrow \alpha_1 \vee \alpha_2 \in L_{IS}$ and $SEM_{IS}(\alpha_1 \vee \alpha_2) = SEM_{IS}(\alpha_1) \cup SEM_{IS}(\alpha_2)$.

Definition 6.2 (*Granule of IS*) Given an information system $IS = (U, A)$. A granule of *IS* is defined by

(*name*, *content*) = $(\alpha, SEM_{IS}(\alpha))$, where $\alpha \in L_{IS}$.

Remark 6.1 Given an information system $IS = (U, A)$. A partition of U over an attribute $a \in A$ is defined by a set of elementary granules $G_a = \{SEM_{IS}(a, v) : v \in V_a\}$.

---

[1]The approach introduced in this chapter can also be applied with no changes to a framework that uses a covering of the universe to form granules.

[2]In this approach the equality relation in the construction of conditions is used. The approach can easily be extended to a case where the conditions are also constructed by applying equality relations and a membership relation.

[3]The notation $SEM_{IS}((a, v))$ is simplified by writing $SEM_{IS}(a, v)$.

Example 6.1 The information system from Example 2.2 can be transformed into the following granular counterpart[4]:

$U\backslash A$	Name	Age	Gender	Income	Class
1	$(AS, \{1\})$	$(36, \{1\})$	$(m, \{1, 6, 7\})$	$(1500, \{1\})$	$(y, \{1, 2, 4, 5, 6\})$
2	$(TJ, \{2\})$	$(33, \{2, 7\})$	$(f, \{2, 3, 4, 5\})$	$(2500, \{2, 5\})$	$(y, \{1, 2, 4, 5, 6\})$
3	$(AT, \{3\})$	$(30, \{3, 4\})$	$(f, \{2, 3, 4, 5\})$	$(1800, \{3, 4, 7\})$	$(n, \{3, 7\})$
4	$(SC, \{4\})$	$(30, \{3, 4\})$	$(f, \{2, 3, 4, 5\})$	$(1800, \{3, 4, 7\})$	$(y, \{1, 2, 4, 5, 6\})$
5	$(ES, \{5\})$	$(26, \{5\})$	$(f, \{2, 3, 4, 5\})$	$(2500, \{2, 5\})$	$(y, \{1, 2, 4, 5, 6\})$
6	$(JC, \{6\})$	$(29, \{6\})$	$(m, \{1, 6, 7\})$	$(3000, \{6\})$	$(y, \{1, 2, 4, 5, 6\})$
7	$(MT, \{7\})$	$(33, \{2, 7\})$	$(m, \{1, 6, 7\})$	$(1800, \{3, 4, 7\})$	$(n, \{3, 7\})$

Conjunctions of atomic formulas (i.e. elementary granules) are used to construct information granules.

Definition 6.3 (*Refinement of a set of elementary granules*) Given an information system $IS = (U, A)$. Let G_a, G_b be sets of elementary granules, where $a, b \in A$.
G_a is a refinement of G_b if every elementary granule in G_a is contained in some elementary granule in G_b. In this case we can say that G_a is finer than G_b or G_b is coarser than G_a.

Sets of elementary granules introduced by more than one attribute are defined recursively:
If G_a and G_b are sets of elementary granules, then so is $G_{a,b} = \{g \cap g' \neq \emptyset : g \in G_a, g' \in G_b\}$.
It is easy to observe that $G_{a,b}$ is finer than G_a and G_b. A finer set, in general, includes a higher number of smaller granules. Therefore, it can be viewed as a finer level of granularity of the universe compared with a coarser set.

6.3 Compound Information Systems

This section introduces a compound information system for deriving information granules from relational data.
A compound information system is constructed based on information systems corresponding to database tables. Firstly, two general definitions of the syntax and semantics of the language are introduced.
Let L be a language such that the syntax and semantics of an atomic formula $\alpha \in L$ and its negation $\neg\alpha \in L$ are defined.

---

[4]Symbolic values are abbreviated to their first letters. Granules in the table are presented in a simplified form, e.g. the granule $(30, \{3, 4\})$ from column *age* corresponds to the granule $((age, 30), \{3, 4\})$.

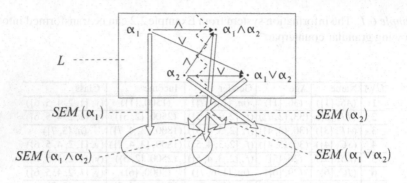

Fig. 6.1 Expansion of language L by the conjunction and disjunction of atomic formulas

Definition 6.4 (*Syntax and semantics of L*) The syntax and semantics of the language L are defined recursively by those of $\alpha \in L$ and $\neg \alpha \in L$, and by

1. $\alpha_1, \alpha_2 \in L \Rightarrow \alpha_1 \wedge \alpha_2 \in L$ and $SEM(\alpha_1 \wedge \alpha_2) = SEM(\alpha_1) \cap SEM(\alpha_2)$;
2. $\alpha_1, \alpha_2 \in L \Rightarrow \alpha_1 \vee \alpha_2 \in L$ and $SEM(\alpha_1 \vee \alpha_2) = SEM(\alpha_1) \cup SEM(\alpha_2)$.

In Fig. 6.1 for any atomic formulas, language L is expanded by the conjunction and disjunction of the formulas (black arrows labeled with $\wedge$ and $\vee$). The semantics of new formulas is constructed based on that of the atomic formulas (white arrows).

Let now $L = L_1 \cup \cdots \cup L_k$ ($k > 1$) be a language such that for each L_i ($1 \le i \le k$) the syntax and semantics are defined.

Definition 6.5 (*Syntax and semantics of L*) The syntax and semantics of the language L are defined recursively by those of each L_i and by the following

1. $\alpha \in L_i \Rightarrow \alpha \in L$ and $SEM(\alpha) = SEM_i(\alpha)$.[5]
2. $\alpha \in L \Rightarrow \neg \alpha \in L$ and $SEM(\neg \alpha) = SEM_i(\neg \alpha)$, where $\alpha \in L_i$;
3. $\alpha_1, \alpha_2 \in L \Rightarrow \alpha_1 \wedge \alpha_2 \in L$ and $SEM(\alpha_1 \wedge \alpha_2) = SEM(\alpha_1) \cap SEM(\alpha_2)$;
4. $\alpha_1, \alpha_2 \in L \Rightarrow \alpha_1 \vee \alpha_2 \in L$ and $SEM(\alpha_1 \vee \alpha_2) = SEM(\alpha_1) \cup SEM(\alpha_2)$.

In Fig. 6.2 for any defined language L_i its every formula α is added to language L (black unlabeled arrows). Language L is expanded by the negation of every previously added formula (the arrow labeled with $\neg$). The negated formulas are de facto taken from L_i. The semantics of added formulas are unchanged (white arrows). Defining the conjunction and disjunction of formulas, we proceed analogously to Fig. 6.1.

The notion of information system will be slightly redefined.

Definition 6.6 (*Information system for database table*) An information system for a database table with the schema $R_i(id, a_1, \ldots, a_m)$ is a pair $IS_i = (U_i, A_i)$, where $U_i = \{x : x \in R_i\}$ and $A_i = \{id, a_1, a_2, \ldots, a_m\}$.[6]

---

[5] SEM_i is the semantics of L_i.

[6] The index (i.e. the relation identifier) is omitted if this does not lead to a confusion.

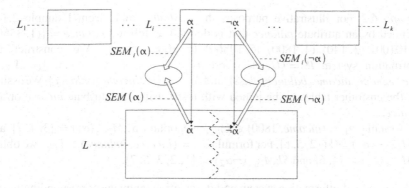

Fig. 6.2 Expansion of language L by the formulas of language L_i

For any database table two types of attributes are distinguished:

1. descriptive attribute—any attribute that can be used to construct a descriptor in a standard attribute-value language;
2. key attribute—any primary/foreign key attribute and any descriptive attribute by which one table can be joined with another table or with itself.

Let $IS = (U, A)$ be an information system of a database table, where $A = A_{des} \cup A_{key}$ and A_{des} (A_{key}) is a set of descriptive (key) attributes.

Definition 6.7 (*Atomic formula in L_{IS}*) An atomic formula in L_{IS} is an expression of either the form

- (a, v), where $a \in A_{des}$ and $v \in V_a$ (*first form*) or
- (a, a'), where $a, a' \in A_{key}$ (*second form*).[7]

Let $L_{IS} = L_{IS_{des}} \cup L_{IS_{key}}$, where $L_{IS_{des}}$ ($L_{IS_{key}}$) consists of formulas of the first (second) form. It is needed to define the syntax and semantics of atomic formulas and their negations of $L_{IS_{des}}$ and $L_{IS_{key}}$, and then apply Definition 6.4 to $L_{IS_{des}}$ and $L_{IS_{key}}$, and Definition 6.5 to L_{IS}.

Definition 6.8 (*Syntax and semantics of $L_{IS} = L_{IS_{des}} \cup L_{IS_{key}}$*) The syntax and semantics of the language L_{IS} are defined recursively by the following, by Definition 6.4 (applied to $L_{IS_{des}}$ and $L_{IS_{key}}$) and by Definition 6.5 (applied to L_{IS})

1. $a \in A_{des}, v \in V_a \Rightarrow (a, v) \in L_{IS_{des}}$ and $SEM_{IS_{des}}(a, v) = \{x \in U : a(x) = v\}$;
2. $\alpha \in L_{IS_{des}} \Rightarrow \neg\alpha \in L_{IS_{des}}$ and $SEM_{IS_{des}}(\neg\alpha) = U\backslash SEM_{IS_{des}}(\alpha)$,
3. $a, a' \in A_{key} \Rightarrow (a, a') \in L_{IS_{key}}$ and $SEM_{IS_{key}}(a, a') = \{x \in U : a(x) = a'(x)\}$;
4. $\alpha \in L_{IS_{key}} \Rightarrow \neg\alpha \in L_{IS_{key}}$ and $SEM_{IS_{key}}(\neg\alpha) = U\backslash SEM_{IS_{key}}(\alpha)$.

---

[7]It is assumed by default that a condition can be constructed based on two key attributes if they are of the same type.

Example 6.2 For illustrative purposes the *customer* table from Example 2.1 is extended by an attribute *balance* that is defined as follows $balance = \{(1, 3550),$ $(2, 40100), (3, 140), (4, 1800), (5, 10860), (6, 3000), (7, 0)\}$. We construct the information system $IS = (U, A)$, where $U = \{1, \ldots, 7\}, A = A_{des} \cup A_{key}, A_{des} = \{age, gender, income, balance, class\}$, and $A_{key} = \{id, income, balance\}$. We assume that the customer table can be joined with itself using the attribute *income* or/and *balance*.

For formula $\alpha_1 = (income, 1800) \in L_{IS_{des}}$ we obtain $SEM_{IS_{des}}(\alpha_1) = \{3, 4, 7\}$ and $SEM_{IS_{des}}(\neg\alpha_1) = \{1, 2, 5, 6\}$. For formula $\alpha_2 = (income, balnace) \in L_{IS_{key}}$ we obtain $SEM_{IS_{key}}(\alpha_2) = \{4, 6\}$ and $SEM_{IS_{key}}(\neg\alpha_2) = \{1, 2, 3, 5, 7\}$.

A compound information system and description language corresponding to two database tables are defined as follows.

Definition 6.9 (*Compound information system $IS_{(i,j)}$*) Let $IS_i = (U_i, A_i)$ and $IS_j = (U_j, A_j)$, where $i \neq j$, be information systems. A compound information system $IS_{(i,j)}$ is defined by[8]

$$IS_{(i,j)} = \times(IS_i, IS_j) = (U_i \times U_j, A_i \cup A_j) \tag{6.1}$$

Definition 6.10 (*Atomic formula in $L_{IS_{(i,j)}}$*) An atomic formula in $L_{IS_{(i,j)}}$ is an expression of either the form

- any atomic formula from L_{IS_i} or L_{IS_j} (*first and second form*) or
- (a, a'), where $a \in (A_i)_{key}$, $a' \in (A_j)_{key}$ (*third form*).[9]

Let $L_{IS_{(i,j)}} = L_{IS_{i\lor j}} \cup L_{IS_{i\land j}}$, where $L_{IS_{i\lor j}}$ consists of formulas from L_{IS_i} and L_{IS_j} (i.e. first and second form formulas) and $L_{IS_{i\land j}}$ consists of formulas of the third form constructed over IS_i and IS_j.

Definition 6.11 (*Syntax and semantics of $L_{IS_{(i,j)}}$*) The syntax and semantics of the language $L_{IS_{(i,j)}}$ are defined recursively by those of L_{IS_i} and L_{IS_j}, by the following, by Definition 6.4 (applied to $L_{IS_{i\lor j}}$ and $L_{IS_{i\land j}}$), and by Definition 6.5 (applied to $L_{IS_{(i,j)}}$)

1. $\alpha \in L_{IS_i} \Rightarrow \alpha \in L_{IS_{i\lor j}}$ and $SEM_{IS_{i\lor j}}(\alpha) = SEM_{IS_i}(\alpha) \times U_j$;
2. $\alpha \in L_{IS_j} \Rightarrow \alpha \in L_{IS_{i\lor j}}$ and $SEM_{IS_{i\lor j}}(\alpha) = U_i \times SEM_{IS_j}(\alpha)$;
3. $\alpha \in L_{IS_{i\lor j}} \Rightarrow \neg\alpha \in L_{IS_{i\lor j}}$ and $SEM_{IS_{i\lor j}}(\neg\alpha) = (U_i \times U_j)\backslash SEM_{IS_{i\lor j}}(\alpha)$;
4. $a \in (A_i)_{key}, a' \in (A_j)_{key} \Rightarrow (a, a') \in L_{IS_{i\land j}}$ and $SEM_{IS_{i\land j}}(a, a') = \{(x, y) \in U_i \times U_j : a(x) = a'(y)\}$;
5. $\alpha \in L_{IS_{i\land j}} \Rightarrow \neg\alpha \in L_{IS_{i\land j}}$ and $SEM_{IS_{i\land j}}(\neg\alpha) = (U_i \times U_j)\backslash SEM_{IS_{i\land j}}(\alpha)$.

---

[8]The intersection of A_i and A_j is empty because all attributes names are distinct from one another, e.g. *customer.id* $\neq$ *purchase.id*.

[9]1. The subset of A_i that consists of all key attributes is denoted by $(A_i)_{key}$. 2. As previously, it is assumed that key attributes are of the same type.

Example 6.3 Consider the information system $IS_{(1,2)} = \times(IS_1, IS_2)$, where IS_1 and IS_2 are constructed respectively based on relations $R_1 = customer$ and $R_2 = purchase$ from Example 2.1.

For formula $\alpha_1 = (age, 1800) \in L_{(IS_1)_{des}}$ we obtain $SEM_{(IS_1)_{des}}(\alpha_1) = \{3, 4, 7\}$ and $SEM_{IS_{1\vee2}}(\alpha_1) = \{3, 4, 7\} \times U_2$.

For formula $\alpha_2 = (R_1.id, R_2.cust\_id) \in L_{IS_{1\wedge2}}$ we obtain $SEM_{IS_{1\wedge2}}(\alpha_2) = \{(1, 1), (1, 2), (2, 3), (2, 4), (3, 8), (4, 5), (4, 6), (6, 7)\}$.

For formula $\alpha_3 = \alpha_1 \wedge \alpha_2 \in L_{IS_{(1,2)}}$ we obtain $SEM_{IS_{(1,2)}}(\alpha_3) = \{(3, 8), (4, 5), (4, 6)\}$.

A compound information system and description language corresponding to m database tables are defined as follows.

Definition 6.12 (*Compound information system $IS_{(1,2,\dots,m)}$*) Let $IS_i = (U_i, A_i)$ be information systems, where $1 \le i \le m$ and $m > 1$ is a fixed number. A compound information system $IS_{(1,2,\dots,m)}$ is defined by

$$IS_{(1,2,\dots,m)} = \times(IS_1, IS_2, \dots, IS_m) = \left(\prod_{i=1}^{m} U_i, \bigcup_{i=1}^{m} A_i\right). \tag{6.2}$$

We will write $IS_{(m)}$ for $IS_{(1,2,\dots,m)}$.

Definition 6.13 (*Syntax and semantics of $L_{IS_{(m)}}$*) The syntax and semantics of the language $L_{IS_{(m)}}$ are defined recursively by those of L_{IS_i} and $L_{IS_{(i,j)}}$ ($1 \le i < j \le m$), by the following, and by Definition 6.4 (applied to $L_{IS_{(m)}}$).

1. $\alpha \in L_{IS_i} \Rightarrow \alpha \in L_{IS_{(m)}}$ and $SEM_{IS_{(m)}}(\alpha) = U_1 \times \cdots \times U_{i-1} \times SEM_{IS_i}(\alpha) \times U_{i+1} \times \cdots \times U_m$;
2. $\alpha \in L_{IS_{(i,j)}} \Rightarrow \alpha \in L_{IS_{(m)}}$ and $SEM_{IS_{(m)}}(\alpha) = \{(x_1, \dots, x_i, \dots, x_j, \dots, x_m) \in \prod_{k=1}^{m} U_k : (x_i, x_j) \in SEM_{IS_{(i,j)}}(\alpha)\}$;
3. $\alpha \in L_{IS_{(m)}} \Rightarrow \neg\alpha \in L_{IS_{(m)}}$ and $SEM_{IS_{(m)}}(\neg\alpha) = (U_1 \times \cdots \times U_m)\backslash SEM_{IS_{(m)}}(\alpha)$.

Since knowledge discovery is focused on selected database tables only, usually one table (i.e. the target table), the semantics of $L_{IS_{(m)}}$ is expanded by the following

1. $\alpha \in L_{IS_{(m)}} \Rightarrow SEM_{IS_{(m)}}^{\pi_i}(\alpha) = \pi_{A_i}(SEM_{IS_{(m)}}(\alpha))$, where $1 \le i \le m$[10];
2. $\alpha \in L_{IS_{(m)}} \Rightarrow SEM_{IS_{(m)}}^{\pi_{i_1, i_2, \dots, i_k}}(\alpha) = \pi_{A_{i_1}, A_{i_2}, \dots, A_{i_k}}(SEM_{IS_{(m)}}(\alpha))$, where $1 \le i_1, i_2, \dots, i_k \le m$ and $k < m$.

Example 6.4 Consider the information system $IS_{(4)} = \times(IS_1, IS_2, IS_3, IS_4)$, where IS_1, IS_2, IS_3, and IS_4 are constructed respectively based on relations $R_1 = customer, R_2 = married\_to, R_3 = purchase$, and $R_4 = product$ from Example 2.1.

---

[10]$\pi_A(\bullet)$ is understood as a projection over the attributes from A.

For formula $\alpha_1 = (age, 1800) \in L_{IS_1}$ we obtain $SEM_{IS_{(4)}}(\alpha_1) = \{3, 4, 7\} \times U_2 \times U_3 \times U_4$.

For formula $\alpha_2 = (R_1.id, R_3.cust\_id) \in L_{IS_{(1,3)}}$ we obtain $SEM_{IS_{(4)}}(\alpha_2) = \{1\} \times U_2 \times \{1, 2\} \times U_4 \cup \{2\} \times U_2 \times \{3, 4\} \times U_4 \cup \{3\} \times U_2 \times \{8\} \times U_4 \cup \{4\} \times U_2 \times \{5, 6\} \times U_4 \cup \{6\} \times U_2 \times \{7\} \times U_4$.

For formula $\alpha_3 = \alpha_1 \wedge \alpha_2 \in L_{IS_{(1,3)}}$ we obtain $SEM_{IS_{(4)}}(\alpha_3) = \{3\} \times U_2 \times \{8\} \cup \{4\} \times U_2 \times \{5, 6\} \times U_4$, $SEM_{IS_{(4)}}^{\pi_{1,3}}(\alpha_3) = \{(3, 8), (4, 5), (4, 6)\}$ and $SEM_{IS_{(4)}}^{\pi_1}(\alpha_3) = \{3, 4\}$.

The compound information system makes it possible to store objects that belong to relations defined extensionally. However, relations defined intensionally can also be expressed in the extended propositional language. Suppose that we are interested in which pairs of persons are married couples, without indicating the order as it is done in relation $R = married\_to$. We can define a new relation $R' = marriage$ with the schema $marriage(\textbf{id}, cust\_id_1, cust\_id_2)$ by means of the rule
$((R'.cust\_id_1, R.cust\_id_1) \wedge (R'.cust\_id_2, R.cust\_id_2)) \vee$
$((R'.cust\_id_1, R.cust\_id_2) \wedge (R'.cust\_id_2, R.cust\_id_1)) \rightarrow (R'.id, R'.id)$.[11]

The following introduces an extension of an attribute-value language to express granules to be derived from relational data.

Definition 6.14 (*Extended attribute-value language*) Given an information system $IS_{(m)} = \times(IS_1, IS_2, \ldots, IS_m)$, where $IS_i = (U_i, A_i)$ $(1 \leq i \leq m)$. An extended attribute-value language is an attribute-value language that includes features of the following forms, and their negations

1. (a, v), where $a \in (A_i)_{des}$ and $v \in V_a$;
2. (a, a'), where $a, a' \in (A_i)_{key}$;
3. (a, a'), where $a \in (A_i)_{key}$, $a' \in (A_j)_{key}$ and $i \neq j$.

The compound information system is a logical representation of a relational database devoted for pattern discovery. Relational data can be represented physically by information systems corresponding to database tables and by the definition of composition of the systems. From the practical point of view, the universe constructed as the Cartesian product of particular universes is too large to be stored. To limit the universe, only possible joins between particular universes are taken into account.

6.4 Constrained Compound Information Systems

This section introduces a constrained compound information system. Constraints used in this system show how particular universes can be connected with one another. To construct constraints one can adapt the relational database notion, i.e. inner or outer join. In this study, left outer join defined by third form formulas is used.

---

[11] The rule conclusion is a trivial formula and means that an object which satisfies the formula belongs to the relation.

Definition 6.15 (*Left outer join on third form formula*) Let $IS_i = (U_i, A_i)$ and $IS_j = (U_j, A_j)$ be information systems. Let also $\theta \in L_{IS_{i \wedge j}}$ be a third form atomic formula. A left outer join on θ is defined by

$$U_i \bowtie_\theta U_j = SEM_{IS_{i \wedge j}}(\theta) \cup \{(x, null) : x \in U_i \setminus SEM_{IS_{i \wedge j}}^{\pi_i}(\theta)\}. \tag{6.3}$$

This definition guarantees that each $x \in U_i$ is included in $U_i \bowtie_\theta U_j$.

Definition 6.16 (*Left outer join on disjunction of third form formulas*) Let $\Theta = \{\theta_1, \theta_2, \ldots \theta_n\}$ be a set of joins of information systems $IS_i = (U_i, A_i)$ and $IS_j = (U_j, A_j)$, i.e. a set of third form atomic formulas such that $\underset{\theta \in \Theta}{\forall} \theta \in L_{IS_{i \wedge j}}$. A left outer join on a disjunction of all the conditions from Θ is defined by

$$U_i \bowtie_\Theta U_j = \bigcup_{\theta \in \Theta} SEM_{IS_{i \wedge j}}(\theta) \cup \{(x, null) : x \in U_i \setminus \bigcup_{\theta \in \Theta} SEM_{IS_{i \wedge j}}^{\pi_i}(\theta)\}. \tag{6.4}$$

This definition guarantees that $(x, null)$ is added to $U_i \bowtie_\Theta U_j$ if and only if $x \in U_i$ is not in any relation defined by $\theta \in \Theta$ with any object from U_j.

The constrained compound information system and the description language corresponding to two database tables are defined as follows.

Definition 6.17 (*Constrained compound information system $IS_{(i,j)}^\Theta$*) Let $IS_i = (U_i, A_i)$ and $IS_j = (U_j, A_j)$ be information systems. Let also $\Theta = \{\theta_1, \theta_2, \ldots \theta_n\}$ be a set of joins of IS_i and IS_j. A constrained compound information system $IS_{(i,j)}^\Theta$ is defined by

$$IS_{(i,j)}^\Theta = \bowtie_\Theta (IS_i, IS_j) = (U_i \bowtie_\Theta U_j, A_i \cup A_j). \tag{6.5}$$

The language $L_{IS_{(i,j)}^\Theta} = L_{IS_{i \vee j}^\Theta} \cup L_{IS_{i \wedge j}^\Theta}$ is defined analogously to that associated with the compound information system. The syntax and semantics of $L_{IS_{(i,j)}^\Theta}$ are defined in the same way as in Definition 6.11. It is enough to replace $IS_{(i,j)}$, $IS_{i \vee j}$, $IS_{i \wedge j}$, and the $\times$ operation with $IS_{(i,j)}^\Theta$, $IS_{i \vee j}^\Theta$, $IS_{i \wedge j}^\Theta$, and the $\bowtie_\Theta$ operation, respectively.

Example 6.5 1. Consider the information system $IS_{(1,2)}^\Theta = \bowtie_{\Theta_1}(IS_1, IS_2)$, where IS_1 and IS_2 are constructed respectively based on relations $R_1 = customer$ and $R_2 = purchase$ from Example 2.1. The set of joins is defined as follows $\Theta_1 = \{\theta\}$, where $\theta = (R_1.id, R_2.cust\_id)$.
    We have $U_1 \bowtie_{\Theta_1} U_2 = \{(1, 1), (1, 2), (2, 3), (2, 4), (3, 8), (4, 5), (4, 6), (5, null), (6, 7), (7, null)\}$. We obtain $\theta \in L_{IS_{(1,2)}^{\Theta_1}}$ and $SEM_{IS_{(1,2)}^{\Theta_1}}(\theta) = \{(1, 1), (1, 2), (2, 3), (2, 4), (3, 8), (4, 5), (4, 6), (6, 7)\}$.
2. Consider also the information system $IS_{(1,3)}^\Theta = \bowtie_{\Theta_2}(IS_1, IS_3)$, where IS_3 is constructed based on relation $R_3 = married\_to$ and $\Theta_2 = \{\theta_1, \theta_2\}$, $\theta_1 = (R_1.id, R_3.cust\_Id_1)$, $\theta_2 = (R_1.id, R_3.cust\_id_2)$.
    We have $U_1 \bowtie_{\Theta_2} U_3 = \{(1, 1), (2, null), (3, 3), (4, 2), (5, 1), (6, 2), (7, 3)\}$. We obtain $\theta_1 \in L_{IS_{(1,3)}^{\Theta_2}}$ and $SEM_{IS_{(1,3)}^{\Theta_2}}(\theta_1) = \{(3, 3), (5, 1), (6, 2)\}$.

The constrained compound information system and the description language corresponding to m database tables are defined as follows.

Definition 6.18 (*Constrained compound information system* $IS^\Theta_{(1,2,\ldots,m)}$) Let $IS_i = (U_i, A_i)$ be information systems, where $1 \le i \le m$ and $m > 1$ is a fixed number, and $\Theta = \{\theta_1, \theta_2, \ldots, \theta_k\}$ be a set of joins such that $\underset{1<j\le m}{\forall} \underset{i<j}{\exists} \; U_i \bowtie_\Theta U_j \ne \emptyset$ (each information system joins with some earlier considered system).
A constrained compound information system $IS^\Theta_{(1,2,\ldots,m)}$ is defined by

$$IS^\Theta_{(1,2,\ldots,m)} = \bowtie_\Theta (IS_1, IS_2, \ldots, IS_m) = (U_1 \bowtie_\Theta U_2 \bowtie_\Theta \cdots \bowtie_\Theta U_m, \bigcup_{i=1}^m A_i).$$
(6.6)

As previously, we will write $IS^\Theta_{(m)}$ for $IS^\Theta_{(1,2,\ldots,m)}$.
The syntax and semantics of $L_{IS^\Theta_{(m)}}$ are defined in the same way as in Definition 6.13.
It is enough to replace $IS_{(i,j)}$, $IS_{(m)}$, and the $\times$ operation with $IS^\Theta_{(i,j)}$, $IS^\Theta_{(m)}$, and the $\bowtie_\Theta$ operation, respectively.
The semantics of $L_{IS^\Theta_{(m)}}$ is expanded by the following

1. $\alpha \in L_{IS^\Theta_{(m)}} \Rightarrow SEM^{\pi_{A_i}}_{IS^\Theta_{(m)}}(\alpha) = \pi_{A_i}(SEM_{IS^\Theta_{(m)}}(\alpha))$, where $1 \le i \le m$;

2. $\alpha \in L_{IS^\Theta_{(m)}} \Rightarrow SEM^{\pi_{i_1,i_2,\ldots,i_k}}_{IS^\Theta_{(m)}}(\alpha) = \pi_{A_{i_1},A_{i_2},\ldots,A_{i_k}}(SEM_{IS^\Theta_{(m)}}(\alpha))$, where
$1 \le i_1, i_2, \ldots, i_k \le m$ and $k < m$.

Example 6.6 Consider the information system $IS^\Theta_{(4)} = \bowtie_\Theta (IS_1, IS_2, IS_3, IS_4)$, where IS_1, IS_2, IS_3, and IS_4 correspond respectively to relations $R_1 = customer$, $R_2 = married\_to$, $R_3 = purchase$, and $R_4 = product$ from Example 2.1, and $\Theta = \{(R_1.id, R_2.cust\_id_1), (R_1.id, R_2.cust\_id_2), (R_1.id, R_3.cust\_id), (R_3.prod\_id, R_4.id)\}$.
We have $U_1 \bowtie_\Theta U_2 \bowtie_\Theta U_3 \bowtie_\Theta U_4 = \{(1, 1, 1, 1), (1, 1, 2, 3), (2, null, 3, 1), (2, null, 4, 3), (3, 3, 8, 4), (4, 2, 5, 6), (4, 2, 6, 2), (5, 1, null, null), (6, 2, 7, 5), (7, 3, null, null)\}$.
The universe consists of 10 elements and is over 134 times smaller than that constructed using the Cartesian product.
For formula $\alpha_1 = (age, 1800) \in L_{IS_1}$ we obtain $SEM_{IS^\Theta_{(4)}}(\alpha_1) = \{(3, 3, 8, 4), (4, 2, 5, 6), (4, 2, 6, 2), (7, 3, null, null)\}$.
For formula $\alpha_2 = (R_1.id, R_3.cust\_id) \in L_{IS_{(1,3)}}$ we obtain $SEM_{IS^\Theta_{(4)}}(\alpha_2) = \{(1, 1, 1, 1), (1, 1, 2, 3), (2, null, 3, 1), (2, null, 4, 3), (3, 3, 8, 4), (4, 2, 5, 6), (4, 2, 6, 2), (6, 2, 7, 5)\}$.
For formula $\alpha_3 = \alpha_1 \wedge \alpha_2 \in L_{IS_{(1,3)}}$ we obtain $SEM_{IS^\Theta_{(4)}}(\alpha_2) = \{(3, 3, 8, 4), (4, 2, 5, 6), (4, 2, 6, 2)\}$, $SEM^{\pi_{1,3}}_{IS^\Theta_{(4)}}(\alpha_3) = \{(3, 8), (4, 5), (4, 6)\}$, and $SEM^{\pi_1}_{IS^\Theta_{(4)}}(\alpha_3) = \{3, 4\}$.

Like in the case of the compound information system, relational data is represented physically by particular information systems and the constraints (i.e. joins) among the systems. However, for databases relatively small the constrained compound information system can be used as both a logical and physical representation.

6.5 Consistency and Completeness of Granule Description Languages

This section provides a formal evaluation of the approach.

Firstly, consistency and completeness of the languages defined in this work is investigated.

Definition 6.19 (*Consistency and completeness of a language L_{IS}*) A language L_{IS}, where $IS = (U, A)$ is an information system, is consistent and complete if and only if for any formula $\alpha \in L_{IS}$ the following hold

$$SEM_{IS}(\alpha) \cap SEM_{IS}(\neg \alpha) = \emptyset \ (consistency) \tag{6.7}$$

$$SEM_{IS}(\alpha) \cup SEM_{IS}(\neg \alpha) = U \ (completeness) \tag{6.8}$$

For compound information systems we obtain.

Proposition 6.1 [12] *The following hold:*

1. *A language $L_{IS} = L_{IS_{des}} \cup L_{IS_{key}}$ is consistent and complete.*
2. *A language $L_{IS_{(i,j)}}$, where $i \neq j$, is consistent and complete.*
3. *A language $L_{IS_{(m)}}$, where $m > 1$ is a fixed number, is consistent and complete.*

Consistency is not satisfied for a language with the expanded semantics. One of the reasons that this does not hold is the database structure. Namely, if there is one-to-many relationship from table T to table T', then formulas constructed over the two tables and with respect to T are, in general, not consistent regardless of the language.

Example 6.7 Consider the information system $IS_{(1,2)} = \times(IS_1, IS_2)$, where IS_1 and IS_2 correspond respectively to relations $R_1 = customer$ and $R_2 = purchase$ from Example 2.1. Let $\alpha = (R_1.id, R_2.cust\_id) \wedge (R_2.amount, 1)$. We have $\neg \alpha = \neg(R_1.id, R_2.cust\_id) \vee \neg(R_2.amount, 1)$. We obtain $(1, 1) \in SEM_{IS_{(1,2)}}(\alpha)$ and $(1, 2) \in SEM_{IS_{(1,2)}}(\neg \alpha)$. More precisely, $(1, 2) \in SEM_{IS_{(1,2)}}((R_1.id, R_2.cust\_id) \wedge \neg(R_2.amount, 1))$. Hence, $1 \in SEM_{IS_{(1,2)}}^{\pi_1}(\alpha) \cap SEM_{IS_{(1,2)}}^{\pi_1}(\neg \alpha)$.

We say that a language is partially consistent if its each formula is consistent or it is inconsistent due to the database structure. Other inconsistencies are caused by the definition of a granule description language or by the construction of the universe.

Example 6.8 Consider the information system $IS_{(1,2)} = \times(IS_1, IS_2)$ where IS_1 and IS_2 correspond respectively to relations $R_1 = customer$ and $R_2 = married\_to$ from Example 2.1. Take the formula $\alpha = (R_1.id, R_2.cust\_id_1)$.

---

[12]Proofs of the propositions formulated in this chapter can be found in [41].

We obtain $SEM_{IS_{(1,2)}}(\alpha) = \{(3,3),(5,1),(6,2)\}$ and $SEM_{IS_{(1,2)}}(\neg\alpha) = \{(1,1),(1,2),$
$(1,3),(2,1),(2,2),(2,3),(3,1),(3,2),(4,1),(4,2),(4,3),(5,2),(5,3),(6,1),$
$(6,3),(7,1),(7,2),(7,3)\}$. Hence, $SEM_{IS_{(1,2)}}(\alpha) \cap SEM_{IS_{(1,2)}}(\neg\alpha) = \emptyset$.
We also obtain $SEM_{IS_{(1,2)}}^{\pi_1}(\alpha) = \{3,5,6\}$ and $SEM_{IS_{(1,2)}}^{\pi_1}(\neg\alpha) = U_1$. Hence, $SEM_{IS_{(1,2)}}^{\pi_1}$
$(\alpha) \cap SEM_{IS_{(1,2)}}^{\pi_1}(\neg\alpha) \neq \emptyset$. The above inconsistency is due to the construction of the
universe.

Proposition 6.2 *The following hold:*

1. *A language $L_{IS_{(i,j)}}$ with the expanded semantics is complete.*
2. *A language $L_{IS_{(m)}}$ with the expanded semantics is complete.*

We now examine constrained compound information systems.

Proposition 6.3 *Languages $L_{IS_{(i,j)}^\Theta}$, where $i \neq j$, and $L_{IS_{(m)}}^\Theta$, where $m > 1$ is a fixed
number, are consistent and complete.*

Proposition 6.4 *A language $L_{IS_{(i,j)}^\Theta}$ with the expanded semantics, where $i \neq j$, is
partially consistent if and only if* $\underset{\theta,\theta'\in\Theta,\theta\neq\theta'}{\forall} SEM_{IS_{(i,j)}^\Theta}^{\pi_i}(\theta) \cap SEM_{IS_{(i,j)}^\Theta}^{\pi_i}(\theta') = \emptyset.$

One can note that for a typical database (i.e. for any two tables at most one
relationship is specified) its language is partially consistent.

Example 6.9 1. Consider the information system $IS_{(1,2)}^\Theta = \bowtie_\Theta (IS_1, IS_2)$, where
IS_1 and IS_2 correspond respectively to relations $R_1 = customer$ and $R_2 =$
$married\_to$ from Example 2.1, and $\Theta = \{\theta_1, \theta_2\}$, $\theta_1 = (R_1.id, R_2.cust\_id_1)$,
$\theta_2 = (R_1.id, R_2.cust\_id_2)$.
The universe is $U_1 \bowtie_\Theta U_2 = \{(1,1),(2,null),(3,3),(4,2),(5,1),(6,2),$
$(7,3)\}$.
We have $SEM_{IS_{(1,2)}^\Theta}^{\pi_1}(\theta_1) \cap SEM_{IS_{(1,2)}^\Theta}^{\pi_1}(\theta_2) = \{3,5,6\} \cap \{1,4,7\} = \emptyset$. We have two
possible third form atomic formulas $\alpha_1 = \theta_1$ and $\alpha_2 = \theta_2$. We obtain $SEM_{IS_{(1,2)}^\Theta}^{\pi_1}(\alpha_1)$
$\cap SEM_{IS_{(1,2)}^\Theta}^{\pi_1}(\neg\alpha_1) = \{3,5,6\} \cap \{2,1,4,7\} = \emptyset$ and $SEM_{IS_1^\Theta}(\alpha_2) \cap SEM_{IS_1^\Theta}(\neg$
$\alpha_2) = \{1,4,7\} \cap \{2,3,5,6\} = \emptyset$. Therefore, the language is partially
consistent.
2. Take the same relations, except that the second relation is the target one. Consider
therefore the information system $IS_{(2,1)}^{\Theta'} = \bowtie_{\Theta'} (IS_2, IS_1)$, where $\Theta' = \{\theta_1', \theta_2'\}$
and $\theta_1' = (R_2.cust\_id_1, R_1.id)$, $\theta_2' = (R_2.cust\_id_2, R_1.id)$.
The universe is $U_2 \bowtie_\Theta U_1 = \{(1,1),(1,5),(2,4),(2,6),(3,3),(3,7)\}$.
We have $SEM_{IS_{(2,1)}^{\Theta'}}^{\pi_2}(\theta_1') \cap SEM_{IS_{(2,1)}^{\Theta'}}^{\pi_2}(\theta_2') = \{1,2,3\} \cap \{1,2,3\} = \{1,2,3\}$.
We have two possible third form atomic formulas $\alpha_1' = \theta_1'$ and $\alpha_2' = \theta_2'$. We
obtain $SEM_{IS_{(2,1)}^{\Theta'}}^{\pi_2}(\alpha_i') \cap SEM_{IS_{(2,1)}^{\Theta'}}^{\pi_2}(\neg\alpha_i') = \{1,2,3\} \cap \{1,2,3\} = \{1,2,3\}$, where
$i = 1, 2$. Therefore, the language is not is partially consistent.

For the database from the second case of the above example we can only consider
positive formulas. To make negative formulas allowed, we can use sublanguages
defined by particular formulas.

By Proposition 6.4 we obtain

Corollary 6.1 *The sublanguage $L_{IS^\theta_{(i,j)}}$ of the language $L_{IS^\Theta_{(i,j)}}$ with the expanded semantics, where $\theta \in \Theta$ and $i \neq j$, is partially consistent.*

Example 6.10 For the information system $IS^{\Theta'}_{(2,1)}$ from the previous example we use the sublanguages $L_{IS^\theta_{(2,1)}}$ and $L_{IS^{\theta'}_{(2,1)}}$.
We have the following subuniverses $U_2 \bowtie_\theta U_1 = \{(1, 5), (2, 6), (3, 3)\}$ and $U_2 \bowtie_{\theta'} U_1 = \{(1, 1), (2, 4), (3, 7)\}$. We obtain $SEM^{\pi_2}_{IS^\theta_{(2,1)}}(\alpha'_1) \cap SEM^{\pi_2}_{IS^\theta_{(2,1)}}(\neg\alpha'_1) = \{1, 2, 3\} \cap \emptyset = \emptyset$ and $SEM^{\pi_2}_{IS^{\theta'}_{(2,1)}}(\alpha'_2) \cap SEM^{\pi_2}_{IS^{\theta'}_{(2,1)}}(\neg\alpha'_2) = \{1, 2, 3\} \cap \emptyset = \emptyset$.

Proposition 6.5 *A language $L_{IS^\Theta_{(i,j)}}$ with the expanded semantics, where $i \neq j$, is complete.*

Proposition 6.6 *A language $L_{IS^\Theta_{(m)}}$ with the expanded semantics, where $m > 1$ is a fixed number, is*

1. *partially consistent if and only if $\displaystyle\mathop{\forall}_{IS^\Theta_{(i,j)}} \mathop{\forall}_{\theta,\theta' \in \Theta} SEM^{\pi_i}_{IS^\Theta_{(i,j)}}(\theta) \cap SEM^{\pi_i}_{IS^\Theta_{(i,j)}}(\theta') = \emptyset.$*

2. *complete.*

6.6 Conclusions

This chapter has developed a granular computing based framework for analyzing and processing data stored in a relational structure. In this framework, data is placed in a (constrained) compound information system, and relational information granules are constructed using an expanded language for granule description. Furthermore, an attribute-value language has been extended to enable expressing relational patterns. In comparison with a relational language, this one has a simpler syntax, but its expressiveness is not limited. Relational information granules are the basis for discovering patterns such as frequent patterns, association rules, and classification rules.

Chapter 7
From Granular-Data Mining Framework to Its Relational Version

7.1 Introduction

Mining data stored in a relational structure [25] rather than in a flat one is a more challenge task. Such data is distributed over multiple tables, and complex relationships among objects of the database can occur. Nevertheless, many algorithms (e.g., [13, 70, 77]) developed for mining propositional data have been upgraded to a relational case. The idea underlying this approach is to preserve as many features of the algorithm to be upgraded as possible. Therefore, only notions specific for relational data are extended. The most important benefit of this approach is a possibility to use all knowledge and experience related to the development and application of standard data mining algorithms.

The goal of this chapter is to provide a general framework for mining relational data [40]. This is an upgrade of a granular computing based data mining framework to a relational case. The general outline of the process of upgrading is inspired by the methodology for extending attribute-value algorithms [95] (details are given in Sect. 7.2). In the methodology used in this chapter, we start with introducing a general granular computing based framework for mining data. It is constructed on the basis of definitions introduced in [83, 89]. Next, we employ a relational extension of an information system to store data and that of an attribute-value language to express patterns. Subsequently, we define a procedure for translating patterns expressed in the extended attribute-value language into a relational language. Finally, we examine the problem of limiting the search space for discovering patterns.

The remaining of the chapter is organized as follows. Section 7.2 restates the methodology for upgrading a standard data mining algorithm to a relational case. Section 7.3 constructs a framework for a relational extension of a standard granular computing approach. Section 7.4 assesses the methodology complexity. Section 7.5 provides concluding remarks.

© Springer International Publishing AG 2017   65
P. Hońko, *Granular-Relational Data Mining*, Studies in Computational
Intelligence 702, DOI 10.1007/978-3-319-52751-2_7

7.2   Relational Extension of a Standard Data Mining Algorithm

Many of first algorithms for mining relational data were developed based on algorithms devoted to propositional data. The task of upgrading a standard data mining algorithm to a relational case is not trivial and requires much attention. An upgraded algorithm should preserve as many features of the original algorithm as possible. In other words, only crucial notions, e.g. data and patterns representation, are upgraded. Furthermore, the original algorithm should be a special case of its relational counterpart, i.e. they both should produce the same results for identical propositional data.

A general methodology for upgrading a standard data mining algorithm to a relational case was proposed in [95]. This methodology is set in an inductive logic programming (ILP) [24] environment and includes the following steps.

1. *Identify the propositional learner that best matches the learning task.*
   An algorithm is chosen that is able to execute as many operations needed for a given task as possible.
2. *Use interpretations to represent examples.*
   A relational data representation called interpretations is used to represent both propositional and relational data.
3. *Upgrade the representation of propositional hypotheses by replacing attribute-value tests with relational tests and modify the coverage test accordingly.*
   A relational representation is used for expressing patterns to be derived from the data. The notion of pattern satisfiability is also upgraded.
4. *Use θ-subsumption as the framework for generality.*
   A method called θ-subsumption is used for determining if a given pattern is more general than another.
5. *Use an operator under θ-subsumption. Use that one that corresponds closely to the propositional operator.*
   A specialization or generalization operator is chosen to refine patterns. The choice depends on the method the original algorithm applies to construct patterns, i.e. if the top-down (bottom-up) method is used, then the specialization (generalization) operator is chosen.
6. *Use a declarative bias mechanism to limit the search space.*
   Constraints are imposed on patterns to be discovered in order to limit the search space which is much bigger (even infinite) than for propositional data.
7. *Implementation.*
   The algorithm is implemented taking into account the differences resulting from changing the data structure. Some operations such as attribute discretization or attribute set reduction cannot be used directly and need adaptation for relational data.
8. *Evaluate the implementation on propositional and relational data.*
   The effectiveness of the upgraded algorithm is verified twofold: for propositional data by comparing with the results obtained by the original algorithm; for rela-

tional data by comparing with the results obtained by other relational data mining algorithms or by performing a statistical evaluation.
9. *Add interesting extra features.*
   The algorithm is extended by additional features, especially by those specific for mining relational data, e.g. another method for limiting the search space.

The main advantages of methodology can be described as follows.

1. A possibility to exploit all expertise and heuristic available for propositional algorithms.
2. A clear relationship between the upgraded relational algorithm and its propositional counterpart, resulting in e.g. identical results on identical propositional data.

The above methodology is dedicated to upgrading a concrete standard algorithm. Therefore, the replacement of the algorithm, and the more of the data mining task, may cause considerable changes in other steps of the upgrading process.

7.3 Granular Computing Based Relational Data Mining Framework

This section introduces a general methodology for upgrading the granular computing based data mining framework to a relational case. The methodology is simplified with comparison to that presented in Sect. 7.2. Namely, only the steps of the upgrading process that are independent of the algorithm to be extended and the data mining task to be performed are carried out.

1. Defining relational data representation.
   Propositional data is to be treated as a special case of relational one. Therefore, the basic task is to define a common representation for propositional as well as relational data. A typical solution relies on using or adjusting a standard relational language to express propositional data.
   The approach presented in this chapter applies the inverse solution. We start with a propositional representation (i.e. an information system) and extend it to express relational data (i.e. a connection of information systems, each corresponding to one database table).
2. Defining relational pattern representation.
   The pattern representation should be consistent with the data one. Therefore, the way the data is represented determines the pattern representation. Generally speaking, the same language is used to represent data and patterns to be discovered from it.
   Along with defining relational pattern representation, we need to upgrade pattern satisfiability. Relational patterns, unlike propositional ones, involve multiple tables, therefore checking pattern satisfiability for analyzed objects implies checking conditions that concern objects that reside in other tables and are related to

the analyzed objects.

We use an extended attribute-value language to express patterns to be generated from the data. Such patterns are translatable into a relational language.

In this approach data from all database tables is located in one compound information system which makes it easier to check pattern satisfiability.

3. Upgrading the notion of pattern generality.

A standard task done during pattern generation is the comparison of patterns with respect to their generality. In the propositional case, one pattern is more general than another if all conditions of the first pattern are ones of the other pattern. In the relational case the problem of generality is more complicated since variable based conditions are allowed in pattern construction. Therefore, not only the syntax, as in the propositional case, but also the semantics of patterns should be check to compare their generality.

The propositional form used in the approach makes it possible to compare the generality of patterns based on their syntax only.

4. Upgrading pattern refinement.

During propositional pattern generation, patterns are refined by specialization (adding new conditions) or by generalization (removing conditions). In the relational case, patterns can also be refined in a different way, e.g. by changing the scope of terms occurring in the patterns (replacing variables with constants or vice versa).

In the approach, patterns are refined by adding or removing propositional conditions only.

5. Limiting the search space.

The propositional search space for pattern discovery can be limited by basic constraints such as a list of allowed attributes, a set of allowed values for each attribute, the size of patterns, the size of a set of patterns. Relational search space is significantly bigger because of a multi-relational data representation and of different combinations of variables that can occur in pattern conditions.

The search space is here limited in two steps: by data model constraints (it guarantees that only valid relationships can be used to construct conditions), by an expert constraints (it guarantees that only conditions specific to a given problem can be used).

For the purposes of this work we assume that

- a relation denoted by R_1 corresponds to the target table,
- for each database table there exits an attribute id that is the identifier of the table's objects.

7.3.1 Construction of Conditions

We can construct any condition over relational data using formulas of a language $L_{IS^{\Theta}_{(m)}}$.

Given an information system $IS_{(m)}^{\Theta} = \bowtie_{\Theta}(IS_1, IS_2, \ldots, IS_m)$, where $IS_i = (U_i, A_i)$ is constructed based on relation R_i ($1 \leq i \leq m$).

We construct granules with respect to the target relation, i.e. granules of the form $(\alpha, SEM_{IS_{(m)}^{\Theta}}^{\pi_1}(\alpha))$, where $\alpha \in L_{IS_{(m)}^{\Theta}}$.

1. initial condition[1]

   a. trivial condition (the second form formula)
      A condition is to be satisfied by any object of the target relation R_1.
      We have $id \in A_1 \Rightarrow \alpha = (id, id) \in L_{IS_1}$.
   b. relationship condition (the second form formula)
      A condition is constructed based on R_1 where for two attributes a relationship is defined.
      Let $a, b \in A_1$ be attributes for which a relationship is defined. We have $a, b \in A_1 \Rightarrow \alpha = (a, b) \in L_{IS_1}$.

2. attribute-value condition (the first form formula)
   Given a formula α_1. The condition to be added to α_1 is $(a = v)$, where $a \in A_1, v \in V_a$.
   We have $a \in A_1, v \in V_a \Rightarrow \alpha_2 = (a, v) \in L_{IS_1}$.
3. another relation-based condition (the third form formula)
   A condition to be added to α_1 is constructed based on relation R_j ($1 < j \leq m$). Let $a \in A_1$ and $b \in A_j$ be the attributes by which relations R_1 and R_j are to be joined. We have $a \in A_1, b \in A_j \Rightarrow \alpha_2 = (a, b) \in L_{IS_{(1,j)}^{\Theta}}$.
4. recursive condition (the third form formula)
   A condition to be added to α_1 is another condition constructed based on the target relation R_1. Let $a, b \in A_1$ be different attributes by which relations R_1 is to be joined with itself. To this end, we make a copy of the relation. Let $R_1' = R_1$ be a copy of R_1. We have $a \in A_1, b \in A_1' \Rightarrow \alpha_2 = (a, b) \in L_{IS_{(1,1')}^{\Theta}}$.
5. negated condition

   a. attribute-value condition (the first form formula)
      A condition to be added to α_1 is $(a \neq v)$, where $a \in A_1, v \in V_a$.
      We have $a \in A_1, v \in V_a \Rightarrow \alpha_2 = (a, v) \in L_{IS_1}$ and $\alpha_2 \in L_{IS_1} \Rightarrow \neg \alpha_2 \in L_{IS_1}$.
   b. another relation-based condition
      It is done by default. Namely, the values of attributes for which a relationship is not defined are assumed to be different.
   c. another relation-based condition (the third form formula)

---

[1]An initial condition indicates the target relation and may be omitted if this does not lead to a confusion, e.g. formulas $(R_1.id, R_1.id) \wedge (R_1.id, R_2.a)$ and $(R_1.id, R_2.a)$ generated for target relation R_1 are equivalent.

A condition to be added to α_1 is a negated condition constructed based on relation R_j $(1 < j \leq m)$.

Let $a \in A_1$ and $b \in A_j$ be the attributes by which relations R_1 and R_j are to be joined. We have $a \in A_1, b \in A_j \Rightarrow \alpha_2 = (a, b) \in L_{IS^\Theta_{(1,j)}}$ and $\alpha_2 \in L_{IS^\Theta_{(1,j)}} \Rightarrow \neg\alpha_2 \in L_{IS^\Theta_{(1,j)}}$.

6. complex condition[2]

   a. conjunction of relationship conditions (the second form formula)
      Let S be the set of pairs $(a, b) \in A_1 \times A_1$ of attributes for which relationships are defined. We have $\underset{(a,b)\in S}{\forall}\ a, b \in A_1 \Rightarrow \alpha = \bigwedge_S (a, b) \in L_{IS_1}$.
   b. conjunction of conditions joining another relation
      i. relationship condition (the third form formula)
         Given a formula α_1. Let S be the set of pairs $(a, b) \in A_1 \times A_j$ $(1 < j \leq m)$ of attributes by which relations R_1 and R_j are to be joined. We have $\underset{(a,b)\in S}{\forall}\ a \in A_1, b \in A_j \Rightarrow \alpha_2 = \bigwedge_S (a, b) \in L_{IS^\Theta_{(1,j)}}$.
      ii. negated relationship condition (the third form formula)
         Consider α_2 from the previous point. We have $\alpha_2 \in L_{IS^\Theta_{(1,j)}} \Rightarrow \neg\alpha_2 \in L_{IS^\Theta_{(1,j)}}$.
         One can note that a condition that is a negated conjunction of atomic formulas is not normally allowed for patterns. Therefore, the pattern $\alpha_1 \wedge \neg\alpha_2$ can be transformed into an equivalent set of patterns $\{\alpha_1 \wedge \neg(a, b) : (a, b) \in S\}$. However, form the practical viewpoint its convenient not to replace such a pattern with its equivalent being a set of patterns. The main reason is that such a pattern's condition is equivalent to a single condition expressed in a relational language.

The addition of any condition to non-target relation that occur in the formula is done analogously to the addition of any condition to the target relation.

Example 7.1 Consider the information system $IS^\Theta_{(4)} = \bowtie_\Theta (IS_1, IS_2, IS_3, IS_4)$ constructed based on relations $R_1 = customer$, $R_2 = married\_to$, $R_3 = R_1$, $R_4 = purchase$ from Example 2.1.
An example formula valid in $IS^\Theta_{(4)}$ is $\alpha_1 \wedge \alpha_2 \wedge \alpha_3 \wedge \alpha_4 \wedge \alpha_5$ where

$\alpha_1 = (R_1.id, R_1.id)$—an entity is a customer (initial condition),
$\alpha_2 = (R_1.age, 30)$—the customer is age of 30 (attribute-value condition),
$\alpha_3 = (R_1.id, R_2.cust\_id_1)$—the customer is married (another relation based condition),
$\alpha_4 = (R_2.cust\_id_2, R_3.id)$—the customer's spouse is a customer (recursive condition),

---

[2]A complex condition is understood as a set of simple conditions that are to be added simultaneously.

$\alpha_5 = \neg(R_3.id, R_4.cust\_id_1)$—the spouse has not purchased yet (negated condition).

7.3.2 Expression of Patterns

The construction of patterns using formulas of language $L_{IS_{(m)}^{\Theta}}$ will be shown. We define patterns such as frequent patterns (i.e. itemset), association rules, classification rules. We also discuss pattern construction with respect to the notions of patterns generality and pattern refinement.

Let $IS_{(m)}^{\Theta} = \bowtie_{\Theta}(IS_1, IS_2, \ldots, IS_m)$ be an information system.

Definition 7.1 (*Frequent pattern*)[3]

1. An expression of the form $p = \alpha_1 \wedge \alpha_2 \wedge \cdots \wedge \alpha_m$ is a pattern in $IS_{(m)}^{\Theta}$ with respect to R_1 if $\alpha_1 \in L_{IS_1}$ or there exists $1 < i \leq m$ such that $\alpha_1 \in L_{IS_{1 \wedge i}^{\Theta}}$.
2. The frequency of p is $freq_{IS_1}(p) = \dfrac{|SEM_{IS_{(m)}^{\Theta}}^{\pi_1}(p)|}{|U_1|}$, where $IS_1 = (U_1, A_1)$.
3. p is frequent if $freq_{IS_1}(p) \geq t$, where t is a given threshold.

Before defining an association rule, the syntax and semantics of $L_{IS_{(m)}^{\Theta}}$ will be expanded by (cf. [83])

- $\alpha_1, \alpha_2 \in L_{IS_{(m)}^{\Theta}} \Leftrightarrow (\alpha_1, \alpha_2) \in L_{IS_{(m)}^{\Theta}}$;
  $SEM_{IS_{(m)}^{\Theta}}((\alpha_1, \alpha_2)) = (SEM_{IS_{(m)}^{\Theta}}(\alpha_1), SEM_{IS_{(m)}^{\Theta}}(\alpha_2))$.

Definition 7.2 (*Association rule*)

1. An expression of the form $\alpha \rightarrow \beta$, represented by the granule $(\alpha, \beta) \in L_{IS_{(m)}^{\Theta}}$, is an association rule in $IS_{(m)}^{\Theta}$ with respect to R_1 if α, β are patterns with respect to R_1 in $IS_{(m)}^{\Theta}$ such that α is more general than or equal to β.
2. The frequency of $\alpha \rightarrow \beta$ is $freq_{IS_1}(\alpha \rightarrow \beta) = freq_{IS_1}(\beta)$.
3. The confidence of $\alpha \rightarrow \beta$ is $conf_{IS_1}(\alpha \rightarrow \beta) = \dfrac{freq_{IS_1}(\beta)}{freq_{IS_1}(\alpha)}$.

Example 7.2 Let $IS_{(1,2)}^{\Theta} = \bowtie_{\Theta}(IS_1, IS_2)$ be the information system constructed based on relations $R_1 = customer$ and $R_2 = married\_to$ from Example 2.1.
Consider patterns $p_1 = (R_1.id, R_2.cust\_id_1)$ and $p_2 = (age, 1800) \wedge (R_1.id, R_2.cust\_id_1)$. We obtain $SEM_{IS_{(1,2)}^{\Theta}}^{\pi_1}(p_1) = \{1, 3, 4, 5, 6, 7\}$ and $SEM_{IS_{(m)}^{\Theta}}^{\pi_1}(p_2) = \{3, 4, 7\}$ Hence, the patterns' frequencies are $freq_{IS_1}(p_1) = 6/7$, $freq_{IS_1}(p_1) = 3/7$.
We have that p_1 is more general than p_2, then $r : p_1 \rightarrow p_2$ is an association rule with $freq_{IS_1}(r) = 3/7$ and $conf_{IS_1}(r) = 1/2$.

---

[3] In this subsection the term *pattern* is understood as an itemset.

Definition 7.3 (*Classification rule*)

1. An expression of the form $\alpha \to \beta$,[4] represented by the granule $(\alpha, \beta) \in L_{IS^\Theta_{(m)}}$, is a classification rule in $IS^\Theta_{(m)}$ with respect to R_1 if α is a pattern with respect to R_1 and β is one of the forms

   a. (d, v), where $d \in A_1$ is a decision attribute (i.e., class attribute) and $v \in V_d$;
   b. (id, id), where $id \in A_1$.[5]

2. The accuracy and coverage of $\alpha \to \beta$ are respectively

$$acc_{IS_1}(\alpha \to \beta) = \frac{|SEM^{\pi_1}_{IS_{(m)}}(\alpha \wedge \beta)|}{|SEM^{\pi_1}_{IS^\Theta_{(m)}}(\alpha)|} \text{ and } cov_{IS_1}(\alpha \to \beta) = \frac{|SEM^{\pi_1}_{IS^\Theta_{(m)}}(\alpha \wedge \beta)|}{|SEM^{\pi_1}_{IS^\Theta_{(m)}}(\beta)|}.$$

Conditions of a rule of the second form (1b) are constructed over the sum of all the target relations, but the conclusion is constructed over one of the target relations.

Example 7.3 1. Consider the information system $IS^\Theta_{(1,2)}$ from the previous example. Examine the rule $\alpha \wedge \beta$ where $\alpha = (age, 30) \wedge (R_1.id, R_2.cust\_id_1)$, $\beta = (class, 1)$.
We obtain $SEM^{\pi_1}_{IS_{(1,2)}}(\alpha) = \{3, 4, 7\}$, $SEM^{\pi_1}_{IS^\Theta_{(1,2)}}(\beta) = \{1, 2, 4, 5, 6\}$ and $SEM^{\pi_1}_{IS^\Theta_{(1,2)}}(\alpha \wedge \beta) = \{3\}$. Hence, $acc_{IS_1}(r_1) = 1/3$, $cov_{IS_1}(r_1) = 1/7$.
2. To illustrate the second form of rules we assume that the customers are defined by two separate relations *customer* and $\neg cutomer$ such that the customer of the first (zeroth) class belong to *customer* ($\neg cutomer$). The schema is common for both the relations, and it does not include the class attribute. Let $R_1 = customer$, $\neg R_1 = \neg customer$, $R_1 = R_1 \cup \neg R_1$, $R_2 = married\_to$ and $IS^\Theta_{(1,2)}$. Rule $\alpha \wedge \beta$ is redefined as follows $\alpha = (age, 1800) \wedge (R_1.id, R_2.cust\_id_1)$ and $\beta = (R_1.id, R_1.id)$. We obtain $SEM^{\pi_1}_{IS_{(1,2)}}(\alpha) = \{3, 4, 7\}$, $SEM^{\pi_1}_{IS_{(1,2)}}(\beta) = \{1, 2, 4, 5, 6\}$ and $SEM^{\pi_1}_{IS_{(1,2)}}(\alpha \wedge \beta) = \{3\}$. The remaining calculations are the same as those for r_1.

Now logical constraints for constructing conditions will be introduced. For propositional conditions the following constraint is imposed.

- A condition defined on attribute a can be added to a pattern if no condition defined on a occurs in the pattern.

Construction of relational patterns is limited by the following constraints.

Definition 7.4 (*Constraints on attribute*) A condition defined on attribute a can be added to a pattern if

---

[4]Unlike in Chap. 2, relational classification rules are not written in an inverse form, since they are extensions of propositional classification rules.
[5]The second form is used when the class attribute is not given, and the membership of an object to a class is meant as its belonging to one of target relations.

1. positive condition

   a. the same condition (not taking into account the negation sign) is not added;
   b. no condition defined on a occurs in the pattern (a is a descriptive attribute);
   c. there occurs in the pattern a non-negated relation such that a is an attribute of this relation[6];

2. negative condition
   A negated condition can be added to a pattern if

   a. its positive counterpart can be added to the pattern;
   b. the condition is not a part of a complex condition.

Example 7.4 1. Given an information system $IS_{(4)}^{\Theta} = \bowtie_{\Theta}(IS_1, IS_2, IS_3, IS_4)$ constructed based on relation $R_1 = customer$, $R_2 = married\_to$, $R_3 = purchase$, $R_4 = product$ from Example 2.1.
Consider the pattern $(R_1.id, R_1.id) \wedge (R_1.age, 1800) \wedge \neg(R_1.id, R_3.id)$. We check which of the below conditions that are assumed to be allowed can be added to the pattern.
Positive conditions that can be added: $(R_1.geneder, male)$, $(R_1.id, R_2.cust\_id_1)$;
Positive conditions that cannot be added: $(R_1.id, R_3.id)$ by 1a, $(R_1.age, 27)$ by 1b, $(R_3.amount, 1)$ and $(R_3.prod\_id, R_4.id)$ by 1c.

2. For the purposes of illustration we extend the database by relations *manufacturer* and *supplier* respectively with the schemas *manufacturer* (**id**, *name*), *supplier*(**id**, *name*). We also extend relation *product* by attributes *manu_id* and *supp_id*. These additional relations includes information about the manufacturers and suppliers of products.
Given an information system $IS_{(1,2)}^{\Theta} = \bowtie_{\Theta}(IS_1, IS_2)$, where where IS_1 and IS_2 are constructed based on relations $R_1 = manufacturer$, $R_2 = product$, respectively.
Consider the pattern $(R_1.id, R_1.id) \wedge (R_1.id, R_2.manu\_id)$.
A negative conditions that can be added: $\neg(R_1.id, R_2.supp\_id)$;
Negative conditions that cannot be added: $\neg(R_1.id, R_2.manu\_id)$ by 2a, $\neg(R_1.id, R_2.supp\_id)$ by 2b under assumption that $(R_1.id, R_2.manu\_id) \wedge (R_1.id, R_2.supp\_id)$ is considered as a complex condition.

7.3.2.1 Pattern Generality

The search space for discovering patterns is structured by means the *is more general relation*. The notion of generality for propositional patterns is defined as follows: One pattern is more general than another if all conditions of the first pattern are

---

[6]A non-negated relation is understood as a relation added to a pattern by means of a non-negated condition.

ones of the other pattern. For relational data such a syntax comparison is not sufficient to define generality relation between two patterns. For example, the pattern $product(X_1, X_2, X_3, X_4, X_5)$ is more general than $product(X_1, X_2, X_3, X_4, X_4)$, what can be verified by the semantics comparison only. This problem is solved by applying θ-subsumption as the framework for generality. According to this framework, one pattern is more general than another if there exists a subsumption such that all conditions of the first pattern after applying the substitution are ones of the other pattern. Unfortunately, checking if or not a pattern is more general than another pattern is an NP-complete problem.

In the approach any constraint is expressed by a separate condition, thanks to this, only the syntax comparison is needed to check if one pattern is more general than another. For example, for the patterns $(product.id, product.id)$ and $(product.id, product.id) \wedge (product.manu\_id, product.supp\_id)$, being the equivalents of $product(X_1, X_2, X_3, X_4, X_5)$ and $product(X_1, X_2, X_3, X_4, X_4)$, respectively, it is enough to check if any condition of the first pattern is one of the the the other pattern.

However, some patterns may need to be transformed into its equivalent form before checking which one of them is more general. For example, consider relations $R_1 = customer, R_2 = purchase, R'_2 = R_2$. The pattern $p_1 = (R_1.id, R_2.cust\_id) \wedge (R_2.amount, 1) \wedge (R_2.date, 25/06)$ is less general than the pattern $p_2 = (R_1.id, R_2.cust\_id) \wedge (R_2.amount, 1) \wedge (R_1.id, R'_2.cust\_id) \wedge (R'_2.date, 25/06)$. In order to syntactically show it, the first pattern needs to be transformed as follows $p_1 \Leftrightarrow p_2 \wedge (R_2.id, R'_2.id)$.

7.3.2.2 Pattern Refinement

Propositional patterns are refined by specialization (adding conditions) or generalization (removing conditions).[7]
Relational patterns can be refined by the above as well as by replacing some terms with other terms, thereby producing more or less general conditions. For example, the pattern $product(X_1, X_2, X_3, X_4, X_5)$ can be specialized by replacing X_5 with X_4, i.e. we obtain $product(X_1, X_2, X_3, X_4, X_4)$.
Since any constraint is expressed by a separate condition in the approach, then any refinement is done as in the propositional case, i.e. by adding or removing conditions. For example, the pattern $(product.id, product.id)$ is specialized by adding the condition $(product.manu\_id, product.supp\_id)$, i.e. we obtain the pattern $(product.id, product.id) \wedge (product.manu\_id, product.supp\_id)$.

Furthermore, in a constrained compound information system the direction of refinement is determined. Namely, constraints in such a system are expressed by third form formulas which are not symmetric (i.e. $\underset{(a,a') \in L_{IS^\theta_{i \wedge j}}}{\forall} (a, a') \not\Leftrightarrow (a', a)$ for

---

[7]Additional refinements are possible if we consider also patterns constructed by using conditions of the form (a, V), where V is a set of values an attribute a may take.

any $IS_{i \wedge j}^{\Theta}$). Thanks to this, unnecessary refinement can be eliminated. For example, in the system $IS_{(1,2)}^{\Theta} = \bowtie_{\Theta}(IS_1, IS_2)$, where IS_1 and IS_2 are constructed respectively based on relations $R_1 = customer$ and $R_2 = married\_to$, and $\Theta = \{(R_1.id, R_2.cust\_id_1), (R_1.id, R_2.cust\_id_2)\}$, the pattern $(R_1.id, R_2.cust\_id_1)$ cannot be refined by useless conditions such as $(R_2.cust\_id_1, R_1.id)$ and $(R_2.cust\_id_2, R_1.id)$.

During relational patterns construction a phenomenon called determinacy problem may arise. This occurs when the pattern refinement does not change the pattern's coverage. For example, suppose that each customer of the database from the running example has purchased at least one product. Therefore, refining the pattern $(customer.id, customer.id)$ by adding the condition $(customer.id, purchase.$ $cust\_id)$ does not affect the pattern's coverage. On the other hand, if we are interested in e.g. customers whose have purchased only one piece of some product, we cannot add the condition $(purchase.amount, 1)$ without adding the previous one.

This problem is overcome by applying the lookahead method [49]. Generally speaking, the condition that causes the determinacy problem can be added to the pattern if this is necessary for adding another condition that does not cause the determinacy problem. Such conditions are added in one step as a refinement.

Conditions that cause the determinacy problem can be detected in a constrained compound information system.

Definition 7.5 (*Condition causing determinacy problem*) Given a constrained compound information system $IS_{(m)}^{\Theta} = \bowtie_{\Theta}(IS_1, IS_2, \ldots, IS_m)$, where $IS_i = (U_i, A_i)$ $(1 \leq i \leq m)$. Indeed, Θ is the set of all possible third form conditions.

A condition $\theta \in \Theta$ to be added to a pattern p causes the determinacy problem with respect to IS_i if $SEM_{IS_{(m)}^{\Theta}}^{\pi_i}(p) \subseteq SEM_{IS_{(m)}^{\Theta}}^{\pi_i}(\theta)$.

Example 7.5 Consider relations $R_1 = customer$, $R_2 = married\_to$, $R_3 = purchase$ and $R_4 = product$ from Example 2.1 and the information system $IS_{(4)}^{\Theta} = \bowtie_{\Theta}(IS_1,$ $IS_2, IS_3, IS_4)$, where $\Theta = \{(R_1.id, R_2.cust\_id_1), (R_1.id, R_2.cust\_id_2),$ $(R_1.id, R_3.cust\_id), (R_3.prod\_id, R_4.id)\}$.
We check conditions for determinacy with respect to IS_1 and the pattern $p = (R_1.id, R_1.id) \wedge \neg(R_1.income, 1800)$. Let $U_p = SEM_{IS_{(4)}^{\Theta}}^{\pi_1}(p) = \{1, 2, 5, 6\}$.

For $\theta_1 = (R_1.id, R_2.cust\_id_1)$ we have $U_p \nsubseteq SEM_{IS_4^{\Theta}}^{\pi_1}(\theta_1) = \{3, 5, 6\}$.

For $\theta_2 = (R_1.id, R_2.cust\_id_2)$ we have $U_p \nsubseteq SEM_{IS_4^{\Theta}}^{\pi_1}(\theta_2) = \{1, 4, 7\}$.

For $\theta_3 = (R_1.id, R_3.cust\_id)$ we have $U_p \subseteq SEM_{IS_4^{\Theta}}^{\pi_1}(\theta_3) = \{1, 2, 4, 5, 6\}$.

Hence, only the last condition causes the determinacy problem.

7.3.3 From Granule-Based Patterns to Relational Patterns

A procedure for translating patterns expressed in the extended attribute-value language into a relational language will be defined.

We start with restating the definitions of relational language and its formulas. The relational language (here denoted by L_R) is a restricted first-order language (function-free formulas are allowed). Its alphabet consists of constants, variables, relation names, connectives, and quantifiers.

Definition 7.6 (*Formulas in L_R*) Formulas in L_R are defined recursively by

1. A term t is either a constant or a variable.
2. if R is k-ary relation and $t_1, t_2, \ldots, t_k$ are terms, then $R(t_1, t_2, \ldots, t_k)$ is an atomic formula.
3. if α is an atomic formula and X is a variable, then $\exists_X \alpha$ and $\forall_X \alpha$ are atomic formulas.
4. if α_1 and α_1 are atomic formulas, then so are $\neg\alpha_1, \alpha_1 \wedge \alpha_2, \alpha_1 \vee \alpha_2$.

Given an information system $IS_{(m)}$. Let $\Leftrightarrow_R$ be a binary operator defined by: $\alpha \Leftrightarrow_R \alpha_R$ if and only if $\alpha_R \in L_R$ is a relational equivalent of $\alpha \in L_{IS_{(m)}}$.

Proposition 7.1 [8] *The following holds* $\forall_{\alpha \in L_{IS_{(m)}}} \exists_{\alpha_R \in L_R} \alpha \Leftrightarrow_R \alpha_R$.

Example 7.6 1. trivial initial condition
   We have $(customer.id, customer.id) \Leftrightarrow_R customer(X_1, X_2, X_3, X_4, X_5, X_6)$.
2. relationship initial condition
   Let $product$ be the target relation. We have $(product.manu\_id, product.supp\_id) \Leftrightarrow_R product(X_1, X_2, X_3, X_4, X_4)$.
3. attribute-value condition
   We have $(income, 1800) \Leftrightarrow_R customer(X_1, X_2, X_3, X_4, X_5, X_6) \wedge X_3 = 1800$.
4. another relation-based condition
   We have $(customer.id, customer.id) \wedge (customer.id, purchase.cust\_id) \Leftrightarrow_R customer(X_1, X_2, X_3, X_4, X_5, X_6) \wedge purchase(Y_1, X_1, Y_3, Y_4, Y_5)$.
5. recursive condition
   Given the pattern $(customer.id, cust\_id_1)$. We add the condition $(cust\_id_2, customer'.id)$. We obtain $(customer.id, cust\_id_1) \wedge (cust\_id_2, customer'.id) \Leftrightarrow_R customer(X_1, X_2, X_3, X_4, X_5, X_6) \wedge married\_to(X_1, X_7) \wedge customer'(X_7, X_8, X_9, X_{10}, X_{11}, X_{12})$.
6. negated attribute-value condition
   We have $\neg(age, 30) \Leftrightarrow_R customer(X_1, X_2, X_3, X_4, X_5, X_6) \wedge X_3 \neq 30$.
7. negated relationship condition
   We have $(customer.id, customer.id) \wedge \neg(customer.id, cust\_id_1) \Leftrightarrow_R customer(X_1, X_2, X_3, X_4, X_5, X_6) \wedge \neg married\_to(X_1, Y_2)$.
8. negated conjunction of conditions joining another relation
   Let $manufacturer$ be the target relation. We have $(manufacturer.id, manufacturer.id) \wedge \neg((manufacturer.id, product.manu\_id) \wedge (manufacturer.id, product.supp\_id)) \Leftrightarrow_R manufacturer(X_1, X_2) \wedge \neg product(Y_1, Y_2, Y_3, X_1, X_1)$.

---

[8]Proofs of the propositions formulated in this chapter can be found in [40].

7.3.4 Search Space Constraints

Finally, additional constraints for the search space will be defined. Note that the search space is partially limited by the data model, i.e. the constrained compound information system. Thanks to this, only valid conditions to occur in patterns are allowed. The constraints defined in this subsection make it possible to construct conditions specific to a given problem only.

The propositional search space can be limited by basic constraints such as a list of allowed attributes, a set of allowed values for each attribute, the size of patterns, the size of a set of patterns. Here we consider the constraint that needs to be adapted in the relational case, i.e. the limitation of the set of values an attribute may take.

We start with defining the search space for relational data.

Definition 7.7 (*Relational search space*) The search space $SS_{IS_{(m)}^\Theta}$ of an information system $IS_{(m)}^\Theta$ is defined by $SS_{IS_{(m)}^\Theta} = \{p \in L_{IS_{(m)}^\Theta} : p\ is\ a\ pattern\}$.

The criterion *p is a pattern* differs depending on the data mining task to be solved.

To define a constraint on an attribute the syntax and semantics of L_{IS_i} $(1 \le i \le m)$ is expanded by the following

- $a \in (A_i)_{des}, S \subseteq V_a \Rightarrow const(a, S) = \bigvee_{v \in S} (a, v) \in L_{IS_i};$

$$SEM_{IS_i}(const(a, S)) = \bigcup_{v \in S} SEM_{IS_i}((a, v)).$$

- $a \in (A_i)_{key}, S \subseteq (A_i)_{key} \Rightarrow const(a, S) = \bigvee_{a' \in S} (a, a') \in L_{IS_i};$

$$SEM_{IS_i}(const(a, S)) = \bigcup_{a' \in S} SEM_{IS_i}((a, a')).$$

The syntax and semantics of $L_{IS_{(i,j)}^\Theta}$ $(1 \le i, j \le m)$ is expanded by the following

$$a \in (A_i)_{key}, S \subseteq (A_j)_{key} \Rightarrow const(a, S) = \bigvee_{a' \in S} (a, a') \in L_{IS_{(i,j)}^\Theta};$$
$$SEM_{IS_{(i,j)}^\Theta}(const(a, S)) = \bigcup_{a' \in S} SEM_{IS_{(i,j)}^\Theta}((a, a')).$$

Finally, the syntax and semantics of $L_{IS_{(i,j_1,j_2,...,j_k)}^\Theta}$ is expanded by the following

$$a \in (A_i)_{key}, S = \bigcup_{1 \le l \le k} S_{j_l}, S_{j_l} \subseteq (A_{j_l})_{key} \ (1 \le l \le k) \Rightarrow const(a, S) =$$
$$\bigvee_{1 \le l \le k} const(a, S_{j_l}); SEM_{IS_{(i,j_1,j_2,...,j_k)}^\Theta}(const(a, S)) = \bigcup_{1 \le l \le k} SEM_{IS_{(i,j_l)}^\Theta}(const(a, S_{j_l})).$$

The set of allowed values/attributes for an attribute a can be defined semantically.[9]

1. $S_a^\# = V_a$— constants (default for descriptive attributes);
2. $S_a = \{a'\ is\ a\ key\ attribute : type(a') = type(a)\}$—all attributes of the same type (default for key attributes);
3. $S_a^+ = \{a' \in attr(p) : a \in S_a\}$—attributes of the same type previously used in a pattern p;

---

[9]These constraints corresponds to typical ones used in ILP.

4. $S_a^- = \{a' \notin attr(p) : a \in S_a\}$—new attributes of the same type (i.e. not used in a pattern p);

Example 7.7   1. Consider relation *customer* from Example 2.1 and constraints:
$const(id, \{id\}), const(age, \{v \geq 30 : v \in V_{age}\}), const(income, \{3000\}^c)$.
All the allowed conditions are: $(id, id), (age, 30), (age, 33), (age, 36)$, $(income, v)$, where $v \in V_{income} \setminus \{3000\}$.
2. Consider relations $R_1 = customer$ and $R_2 = purchase$ and constraints:
$const(R_1.id, S_{R_1.id}^-), const(R_2.cust\_id, S_{R_2.cust\_id}^+)$.
Before constructing a pattern we have $S_{R_1.id}^- = \{R_1.id, R_2.cust\_id\}, S_{R_2.cust\_id}^+ = \emptyset$.
For the pattern $(R_1.id, R_2.cust\_id)$ we obtain $S_{R_1.id}^- = \{R_1.id\}, S_{R_2.cust\_id}^+ = \{R_1.id, R_2.cust\_id\}$.

7.4   The Methodology's Complexity

Firstly, the way of the construction of the search space is evaluated. The cost of the formation of all possible conditions based on which patterns are constructed is computed.

Given an information system $IS = (U, A)$, where $A = A_{des} \cup A_{key}$. Let $n = |A|$. Let also $type_{IS}(a) = \{a' \in A_{key} : type(a) = type(a')\}$ be the set of all key attributes from IS with the same type as an attribute a.

1. We assume that the cost of the formation of the condition (a, v), where attribute a and value v are given is 1. The cost of the construction of all conditions for descriptive attributes is

$$T_1(n) = \sum_{a \in A_{des}} \sum_{v \in V_a} 1 \leq |A_{des}|C \leq nC = O(n),$$

where $C = max\{|V_a| : a \in A_{des}\}$ is small and do not depend on the data size since we assume that the data is discretized.
2. We assume that the cost of the formation of the condition (a, a'), where attributes a and a' are given is 1. The cost of the construction of all conditions for key attributes is

$$T_2(n) = \sum_{a \in A_{key}} \sum_{a' \in type_{IS}(a)} 1 \leq |A_{key}|(|A_{key}| - 1) \leq n(n - 1) = O(n^2).$$

In a pessimistic case, we have $type_{IS}(a) = A_{key} \setminus \{a\}$.

Given an information system $IS_{(i,j)}^\Theta = \bowtie_\Theta (IS_i, IS_j)$, where $IS_i = (U_i, A_i)$, $IS_j = (U_j, A_j)$. Let $n_i = |A_i|$ and $n_j = |A_j|$. Let also $\Theta_a \subseteq \Theta$ be the set of conditions constructed based on an attribute $a \in (A_i)_{key}$. The cost of the construction of all relationship conditions for key attributes from IS_1 is

$$T_3(n_i, n_j) = \sum_{a \in (A_i)_{key}} \sum_{\theta \in \Theta_a} 1 \le |(A_i)_{key}| max\{|\Theta_a| : a \in (A_i)_{key}\} \le n_i n_j = O(n_i n_j).$$

Given information system $IS^\Theta_{(1,2,...,m)}$. Let $n_i = |A_i|$ for $i = 1, 2, ..., m$.
Let I be a set of pairs of indexes of information systems to be joined by relationship conditions.
The cost of the construction of all relationship conditions for key attributes from $IS^\Theta_{(1,2,...,m)}$ is

$$T_4(n_1, n_2, ..., n_m) = \sum_{(i,j) \in I} T_3(n_i, n_j) \le \frac{m(m-1)}{2} n_i^{max}(n_j^{max} - 1) =$$

$$C n_i^{max}(n_j^{max} - 1) = O(n_i^{max} n_j^{max}),$$

where $\forall_{n \in \{n_1, n_2, ..., n_m\}} n \le n_i^{max}, n_j^{max}$ and $C = \frac{m(m-1)}{2}$ is the maximal number of pairs of systems to be joined. This value can be big; however it does not depend on the attribute set size and on the algorithm for search space construction since it is determined by the database structure. Therefore, we do not consider the number of information systems as the input data to the algorithm for search space construction. The cost of the construction of all conditions for attributes from $IS^\Theta_{(1,2,...,m)}$ is

$$T_5(n_1, n_2, ..., n_m) = \sum_{i=1}^{m}(T_1(n_i) + T_2(n_i)) + T_4(n_1, n_2, ..., n_m) \le m(T_1(n_i^{max}) +$$

$$T_2(n_i^{max})) + \frac{m(m-1)}{2} T_3(n_i^{max} n_j^{max}) = O(n_i^{max}) + O((n_i^{max})^2) + O(n_i^{max} n_j^{max}) =$$

$$O((n_i^{max})^2),$$

where $\forall_{n \in \{n_1, n_2, ..., n_m\}} n \le n_i^{max}, n_j^{max}$ and $n_i^{max} \ge n_j^{max}$.

Consider a typical relational database: (almost) all tables include descriptive attributes; the number of joins between any two tables is small (usually one join).

For any attribute $a \in A_{(m)} = \bigcup_{i=1}^{m} A_i$ from $IS^\Theta_{(1,2,...,m)} = \bowtie_\Theta (IS_1, IS_2, ..., IS_m)$, where $IS_i = (U_i, A_i)$, we define its domain, denoted $Dom(a)$, as the set of values or/and attributes which a can take. We assume that $\exists_{C \ll n} \forall_{a \in A_{(m)}} |Dom(a)| \le C$.

The cost of the construction of all conditions for attributes from IS is

$$T_1'(n) = \sum_{a \in A} \sum_{d \in Dom(a)} 1 \le |A|C \le nC = O(n).$$

The cost of the construction of all conditions for attributes from $IS^{\Theta}_{(1,2,\dots,m)}$ is

$$T'_5(n_1, n_2, \dots, n_m) = \sum_{i=1}^{m} T'_1(n_i) \le mT'_1(n_i^{max}) = O(n_i^{max}),$$

where $n_i^{max} = max\{n_1, n_2, \dots, n_m\}$.

Now the way of the construction of patterns will be evaluated.

Let $|U| = n$, where $IS = (U, A)$ is an information system.

1. We assume that the cost of checking the condition (a, v) for o, where attribute a, value v and object o are given, is 1. The cost of checking a condition constructed based on a descriptive attribute is

$$T_6(n) = \sum_{o \in U} 1 = O(n).$$

2. The cost of checking a negated condition constructed based on a descriptive attribute (i.e. $\neg(a, v)$) is[10]

$$T'_6(n) \le T_6(n) + 2n = O(n).$$

3. We assume that the cost of checking the condition (a, a') for o, where attributes a and a' and object o are given, is 1. The cost of checking a condition constructed based on an inner key attribute is

$$T_7(n) = \sum_{o \in U} 1 = O(n).$$

4. The cost of checking a negated condition constructed based on an inner key attribute (i.e. $\neg(a, a')$) is

$$T'_7(n) \le T_7(n) + 2n = O(n).$$

Given an information system $IS^{\Theta}_{(i,j)} = \bowtie_{\Theta}(IS_i, IS_j)$, where $IS_i = (U_i, A_i)$, $IS_j = (U_j, A_j)$ and $|\Theta| = l$. Let $n_1 = |A_i|$ and $n_2 = |A_j|$.

We assume that the cost of checking the condition (a, a') for o, o', where attributes $a \in A_{i_{key}}$ and $a' \in A_{j_{key}}$ and objects $o \in U_i, o' \in U_j$ are given, is 1.

---

[10]The cost of the subtraction of the set of objects that satisfy the non-negated condition from the universe equals to or is less than $2n$ since both the sets are assumed to be ordered.

The cost of checking a condition constructed based on an outer key attribute from IS_1 is

$$T_8(n_1, n_2) = \sum_{(o,o') \in U_1 \bowtie_\Theta U_2} 1 \leq lmax\{|SEM_{IS_{1 \wedge 2}}(\theta)| : \theta \in \Theta\} \leq$$

$$lmax\{n_1, n_2\} = O(n_i^{max}),$$

where $n_i^{max} = max\{n_1, n_2\}$ corresponds the maximal number of pairs of objects produced by a left outer join between U_1 and U_2. In a pessimistic case, we have to scan all l joins.

The cost of checking a condition constructed based on any attribute from $IS_{(1,2,\ldots,m)}^\Theta$ is

$$T_9(n_1, n_2, \ldots, n_m) \leq O(n_i^{max}),$$

where $n_i^{max} = max\{n_1, n_2, \ldots, n_m\}$.

Let α be k-ary pattern. The cost of checking all conditions of α is

$$T_{10}(n_1, n_2, \ldots, n_m) \leq kO(n_i^{max}) = O(n_i^{max}).$$

7.5 Conclusions

This chapter has developed a general methodology for relational upgrading a data mining framework defined in a granular computing environment. This environment makes it possible to analyze a given problem at different levels of granularity of relational data.

Unlike its predecessors, the introduced methodology uses not a relational but an (extended) attribute-value language to express data and patterns. Thanks to this, notions such as pattern generality and pattern refinement are unchanged, whereas the following notions are slightly changed: condition construction (except for descriptive conditions, relationship ones are allowed), pattern satisfiability (instead of an object from a single table, a tuple of objects from multiple tables is substituted into a pattern).

The crucial problem of relational data mining is a large search space for discovering patterns. In the approach, this is limited in two steps: by data model constraints (it guarantees that only valid relationships can be used to construct conditions), by an expert constraints (it guarantees that only conditions specific to a given problem can be used).

Chapter 8
Relation-Based Granules

8.1 Introduction

The compound information system (Chap. 6) can be directly mined or can be before-hand transformed into a granular form. The former facilitates the construction of patterns over many tables since the connection among tables are included in the system; however elementary granules that show objects sharing the same features are not contained. The latter, in turn, includes elementary granules (each associated with one table or with two tables to show the connection between them) but the construction of relational patterns over the description language requires granules to be defined so that each of them is associated with all tables under consideration.

To construct a relational data representation that is more coherent and useful for pattern discovery, relation-based granules are introduced in this study [42]. They are formed using relations that join relational information granules with their features. They are used to represent both data and patterns. Relation-based granules are more informative than the granules based on which they are constructed. They include information about how a given granule can alternatively be joined with another from a different information system. The relations used to represent data are fundamental components of patterns. Since the relations express basic features of objects, the process of the generation of patterns can be sped up. Furthermore, the structure of relation-based granules facilities the formation of more advanced conditions. They correspond to those that can be formed by using aggregation functions in relational databases. Therefore, patterns constructed based on such relations show richer knowledge than standard relational patterns.

The remaining of the chapter is organized as follows. Section 8.2 expands the description languages by defining relation-based granules. Section 8.3 shows how relation-based granules can be used for representing both the relational data and patterns. Section 8.4 evaluates the approach's complexity. Section 8.5 provides concluding remarks.

© Springer International Publishing AG 2017
P. Hońko, *Granular-Relational Data Mining*, Studies in Computational
Intelligence 702, DOI 10.1007/978-3-319-52751-2_8

8.2   Construction of Relation-Based Granules

This section introduces an expansion of the description languages by defining
relation-based granules. To distinguish them from those defined in the previous
section, we will call the latter formula-based granules.

A relation is constructed based on a formula to show not only the objects that
satisfy the formula but also the values of attributes that characterize the objects. More
precisely, attribute values and an object are in the relation if and only if the object
satisfies the formula constructed based on the attribute values. Granules constructed
based on the relations include also their characteristics. This information makes it
easy to join granules from different universes.

Let $attr(\alpha)$ denote the set of all attributes used in a formula α.

Definition 8.1 (ε-*relation*) Let $IS_{(m)}$ be a compound information system. A formula-
based relation, called ε-relation, is a relation defined by a formula $\alpha \in L_{IS_{(m)}}$,
denoted by ε_α, with the schema $\varepsilon_\alpha(a_1, \ldots, a_k)$, where $attr(\alpha) \subset \{a_1, \ldots, a_k\}$ and
$\{a_1, \ldots, a_k\} \backslash attr(\alpha)$ consists of key attributes.

Only formulas that express non-negative conditions will be considered in this study.

8.2.1   Information System

Formula-based relations for an information system IS are defined as follows. Let $L_{IS}^\star$
denote an expanded language L_{IS}.

Definition 8.2 (*Syntax and semantics* of $L_{IS}^\star$) The syntax and semantics of a language
$L_{IS}^\star$ are those of L_{IS} expanded recursively by

1. $(a, v) \in L_{IS} \Rightarrow \varepsilon_{(a,v)} \in L_{IS}^\star$ and $SEM_{IS}(\varepsilon_{(a,v)}) = \{v\} \times SEM_{IS}(a, v)$;
2. $(a, v), (a', v') \in L_{IS}, a \neq a' \Rightarrow \varepsilon_{(a,v) \wedge (a',v')} \in L_{IS}^\star$ and $SEM_{IS}(\varepsilon_{(a,v) \wedge (a',v')}) =$
   $\{v\} \times \{v'\} \times SEM_{IS}((a, v) \wedge (a', v'))^1$;
3. $\bigvee_{v \in V_a} (a, v) \in L_{IS} \Rightarrow (a, \cdot) = \bigvee_{v \in V_a} (a, v) \in L_{IS}^\star$ and $SEM_{IS}(a, \cdot) = \bigcup_{v \in V_a}$
   $SEM_{IS}(a, v)$;
4. $(a, \cdot) \in L_{IS}^\star \Rightarrow \varepsilon_{(a,\cdot)} \in L_{IS}^\star$ and $SEM_{IS}(\varepsilon_{(a,\cdot)}) = \bigcup_{v \in V_a} SEM_{IS}(\varepsilon_{(a,v)})$;
5. $(a, \cdot), (a', v') \in L_{IS}^\star \Rightarrow \varepsilon_{(a,\cdot) \wedge (a',v')} \in L_{IS}^\star$ and $SEM_{IS}(\varepsilon_{(a,\cdot) \wedge (a',v')}) = \bigcup_{v \in V_a}$
   $SEM_{IS}(\varepsilon_{(a,v) \wedge (a',v')})$;
6. $(a, v), (a', \cdot) \in L_{IS}^\star \Rightarrow \varepsilon_{(a,v) \wedge (a',\cdot)} \in L_{IS}^\star$ and $SEM_{IS}(\varepsilon_{(a,v) \wedge (a',\cdot)}) = \bigcup_{v' \in V_{a'}}$
   $SEM_{IS}(\varepsilon_{(a,v) \wedge (a',v')})$;
7. $(a, \cdot), (a', \cdot) \in L_{IS}^\star \Rightarrow \varepsilon_{(a,\cdot) \wedge (a',\cdot)} \in L_{IS}^\star$ and $SEM_{IS}(\varepsilon_{(a,\cdot) \wedge (a',\cdot)}) = \bigcup_{v \in V_a}$
   $SEM_{IS}(\varepsilon_{(a,v) \wedge (a',\cdot)}) = \bigcup_{v' \in V_{a'}} SEM_{IS}(\varepsilon_{(a,\cdot) \wedge (a',v')})$.

---

[1]A relation based on disjunction is defined analogously.

Compared with the semantics of formulas, that of the relations provides also information about features the objects share. Furthermore, the relations make it possible to show features of all objects at once (i.e. relations with descriptors of the form $(a, \cdot)$).

The pair $(\varepsilon_\alpha, SEM(\varepsilon_\alpha))$ is viewed as a granule constructed based on the granule $(\alpha, SEM(\alpha))$.

Example 8.1 Given the information system $IS = (U, A)$ constructed based on relation $R = customer$ from Example 2.1.

Consider formulas $\alpha_1 = (age, 30), \alpha_2 = (income, 1800), \alpha_3 = (age, \cdot), \alpha_4 = (income, \cdot) \in L_{IS}^*$ and the relations with the schemas $\varepsilon_{\alpha_1}(age, id), \varepsilon_{\alpha_2}(income, id), \varepsilon_{\alpha_3}(age, id), \varepsilon_{\alpha_4}(income, id)$. The semantics are $SEM_{IS}(\varepsilon_{\alpha_1}) = \{30\} \times \{3, 4\}$, $SEM_{IS}(\varepsilon_{\alpha_2}) = \{1800\} \times \{3, 4, 7\}$, $SEM_{IS}(\varepsilon_{\alpha_3}) = \bigcup\{\{30\} \times \{3, 4\}, \{33\} \times \{2, 7\}, \{26\} \times \{5\}, \{29\} \times \{6\}\}$, $SEM_{IS}(\varepsilon_{\alpha_4}) = \bigcup\{\{1500\} \times \{1\}, \{1800\} \times \{3, 4, 7\}, \{2500\} \times \{2, 5\}, \{3000\} \times \{6\}\}$. We join the formulas and thereby obtain the following relations with the schemas $\varepsilon_{\alpha_1 \wedge \alpha_2}(age, income, id), \varepsilon_{\alpha_3 \wedge \alpha_2}(age, income, id), \varepsilon_{\alpha_1 \wedge \alpha_4}(age, income, id)$, $\varepsilon_{\alpha_3 \wedge \alpha_4}(age, income, id)$. The semantics are $SEM_{IS}(\varepsilon_{\alpha_1 \wedge \alpha_2}) = \{30\} \times \{1800\} \times \{3, 4\},^2 SEM_{IS}(\varepsilon_{\alpha_3 \wedge \alpha_2}) = \bigcup\{\{30\} \times \{1800\} \times \{3, 4\}\}, SEM_{IS}(\varepsilon_{\alpha_1 \wedge \alpha_4}) = \bigcup\{\{30\} \times \{1800\} \times \{3, 4\}\}, SEM_{IS}(\varepsilon_{\alpha_3 \wedge \alpha_4}) = \bigcup\{\{36\} \times \{1500\} \times \{1\}, \{30\} \times \{1800\} \times \{3, 4\}, \{33\} \times \{2500\} \times \{2\}, \{26\} \times \{2500\} \times \{5\}, \{29\} \times \{3000\} \times \{6\}, \{33\} \times \{1800\} \times \{7\}\}$. Each granule constructed based on the above relations shows not only some customers but also features they share, e.g. the granule $(\varepsilon_{\alpha_3 \wedge \alpha_4}, SEM_{IS}(\varepsilon_{\alpha_3 \wedge \alpha_4}))$ shows for each group of customers their ages and incomes.

8.2.2 Compound Information System

Since objects are referenced by their identifiers, the following simplification will be used. Let $IS = (U, A)$ be an information system constructed based on relation $R(id, a_i, \ldots, a_k)$. We assume that $v \in U \equiv v \in V_{id} \wedge \underset{x \in U}{\exists} \, id(x) = v$. An analogous assumption is made for a compound information system.

Formula-based relations for a compound information system $IS_{(i,j)}$ are defined as follows.

Definition 8.3 (*Syntax and semantics of* $L_{IS_{(i,j)}}^*$) The syntax and semantics of a language $L_{IS_{(i,j)}}^*$ are those of $L_{IS_{(i,j)}}$ expanded recursively by those of $L_{IS_1}^*$ and $L_{IS_2}^*$, and by

---

[2]If the $\bigcup$ operation is used for one set only, it means that it is possible to obtain more than one set for the given relation.

1. $(a, v) \in L^{\star}_{IS_i}, (a', v') \in L^{\star}_{IS_j}, v' \in SEM_{IS_i}(a, v) \Rightarrow \varepsilon_{(a,v) \wedge (a',v')} \in L^{\star}_{IS_{(i,j)}}$ and
$SEM_{IS_{(i,j)}}(\varepsilon_{(a,v) \wedge (a',v')}) = \{v\} \times \{v'\} \times SEM_{IS_j}(a', v')$;

2. $(a, v) \in L^{\star}_{IS_i}, (a', v') \in L^{\star}_{IS_j}, v \in SEM_{IS_j}(a', v') \Rightarrow \varepsilon_{(a,v) \wedge (a',v')} \in L^{\star}_{IS_{(i,j)}}$ and
$SEM_{IS_{(i,j)}}(\varepsilon_{(a,v) \wedge (a',v')}) = \{v\} \times SEM_{IS_i}(a, v) \times \{v'\}$;

3. $(a, v) \in L^{\star}_{IS_i}, (a', v') \in L^{\star}_{IS_j}, v = v' \Rightarrow \varepsilon_{(a,v) \wedge (a',v')} \in L^{\star}_{IS_{(i,j)}}$ and $SEM_{IS_{(i,j)}}$
$(\varepsilon_{(a,v) \wedge (a',v')}) = \{v\} \times SEM_{IS_i}(a, v) \times SEM_{IS_i}(a', v') = \{v'\} \times SEM_{IS_i}(a, v) \times$
$SEM_{IS_i}(a', v')^3$;

4. $(a, \cdot) \in L^{\star}_{IS_i}, (a', v') \in L^{\star}_{IS_j} \Rightarrow \varepsilon_{(a,\cdot) \wedge (a',v')} \in L^{\star}_{IS_{(i,j)}}$ and $SEM_{IS_{(i,j)}}$
$(\varepsilon_{(a,\cdot) \wedge (a',v')}) = \bigcup_{v \in V_a} SEM_{IS_{(i,j)}}(\varepsilon_{(a,v) \wedge (a',v')})$;

5. $(a, v) \in L_{IS_i}, (a', \cdot) \in L^{\star}_{IS_j} \Rightarrow \varepsilon_{(a,v) \wedge (a',\cdot)} \in L^{\star}_{IS_{(i,j)}}$ and $SEM_{IS_{(i,j)}}(\varepsilon_{(a,v) \wedge (a',\cdot)}) =$
$\bigcup_{v' \in V_{a'}} SEM_{IS_{(i,j)}}(\varepsilon_{(a,v) \wedge (a',v')})$;

6. $(a, \cdot) \in L^{\star}_{IS_i}, (a', \cdot) \in L^{\star}_{IS_j} \Rightarrow \varepsilon_{(a,\cdot) \wedge (a',\cdot)} \in L^{\star}_{IS_{(i,j)}}$ and $SEM_{IS_{(i,j)}}(\varepsilon_{(a,\cdot) \wedge (a',\cdot)}) =$
$\bigcup_{v \in V_a} SEM_{(i,j)}(\varepsilon_{(a,v) \wedge (a',\cdot)}) = \bigcup_{v' \in V_{a'}} SEM_{(i,j)}(\varepsilon_{(a,\cdot) \wedge (a',v')})$;

7. $\alpha \in L^{\star}_{IS_l} \Rightarrow \varepsilon_\alpha \in L^{\star}_{IS_{(i,j)}}$ and $SEM_{IS_{(i,j)}}(\varepsilon_\alpha) = SEM_{IS_l}(\varepsilon_\alpha)$, where $l = i, j$.

The semantics of the relations defined above show pairs of objects from different universes along with their features. Relations that include a descriptor of the form $(a, \cdot)$ makes it easy to join objects from different universes.

Example 8.2 Given the information system $IS_{(1,2)} = \times (IS_1, IS_2)$, where IS_1 and IS_2 are constructed respectively based on relations $R_1 = customer$ and $R_2 = purchase$ from Example 2.1.
Let $\alpha_1 = (age, 30), \alpha_2 = (cust\_id, \cdot), \alpha_3 = (date, \cdot) \in L^{\star}_{IS_{(1,2)}}$. We consider the following relations with the schemas $\varepsilon_{\alpha_1 \wedge \alpha_2}(age, cust\_id, R_2.id)$,
$\varepsilon_{\alpha_2 \wedge \alpha_3}(cust\_id, date, R_2.id), \varepsilon_{\alpha_1 \wedge \alpha_2 \wedge \alpha_3}(age, cust\_id, date, R_2.id)$. The semantics are
$SEM_{IS_{(1,2)}}(\varepsilon_{\alpha_1 \wedge \alpha_2}) = \bigcup \{\{30\} \times \{3\} \times \{8\}, \{30\} \times \{4\} \times \{5, 6\}\}$,
$SEM_{IS_{(1,2)}}(\varepsilon_{\alpha_2 \wedge \alpha_3}) = \bigcup \{\{1\} \times \{24/06\} \times \{1, 2\}, \{2\} \times \{25/06\} \times \{3\}, \{2\} \times \{26/06\} \times \{4\}, \{3\} \times \{27/06\} \times \{8\}, \{4\} \times \{26/06\} \times \{5, 6\}, \{6\} \times \{27/06\} \times \{7\}\}, SEM_{IS_{(1,2)}}(\varepsilon_{\alpha_1 \wedge \alpha_2 \wedge \alpha_3}) = \bigcup \{\{30\} \times \{3\} \times \{27/06\} \times \{8\}, \{30\} \times \{4\} \times \{26/06\} \times \{5, 6\}\}$.
Each granule constructed based on the above relations shows some customers and their purchases, and also features of at least one of both, e.g. the granule $(\varepsilon_{\alpha_1 \wedge \alpha_2 \wedge \alpha_3}, SEM_{IS}(\varepsilon_{\alpha_1 \wedge \alpha_2 \wedge \alpha_3}))$ shows customers, their ages, their purchases and the purchase dates.
To illustrate case 3 of Definition 8.3 we consider also an information system IS_3 constructed based on relation $R_3 = married\_to$.
Let $\alpha_4 = (cust\_id, \cdot) \wedge ((cust\_id_1, \cdot) \vee (cust\_id_2, \cdot)) \in L^{\star}_{IS_{(2,3)}}$. We have the following relation with the schema $\varepsilon_{\alpha_4}(cust\_id, R_2.id, R_3.cust\_id)$.[4] The semantics is
$SEM_{IS_{(2,3)}}(\varepsilon_{\alpha_4}) = \bigcup \{\{1\} \times \{1, 2\} \times \{1\}, \{4\} \times \{5, 6\} \times \{2\}, \{6\} \times \{7\} \times \{3\}, \{3\} \times \{8\} \times \{3\}\}$.

---

[3] In this case it is assumed that a and a' are of the same type.
[4] The last attribute in $\varepsilon_{\alpha_4}(cust\_id, R_2.id, R_3.cust\_id)$ corresponds to $cust\_id_1$ and $cust\_id_2$.

Formula-based relations for a compound information system $IS_{(m)} = \times(IS_1, IS_2, \ldots, IS_m)$, where $m > 2$, are defined analogously to Definition 8.3.

8.3 Relational Data and Patterns Represented by Relation-Based Granules

This section shows how relation-based granules can be used for representing both the relational data and patterns.

8.3.1 Relational Data Representation

Definition 8.4 (*Relational data representation*) An information system $IS = (U, A)$ is represented by a relation $\varepsilon_{IS} = (\varepsilon_{(a,\cdot)} : a \in A)$.
A database defined by relations $R_1, R_2, \ldots, R_m$ (each R_i corresponds to $IS_i = (U_i, A_i)$) is represented by the set $\{\varepsilon_{IS_1}, \varepsilon_{IS_2}, \ldots \varepsilon_{IS_m}\}$.

To make the data representation more compact, one can omit the relations constructed based on the identifiers since they are included in other relations.

Example 8.3 Let $R_1 = customer, R_2 = purchase, R_3 = product, R_4 = married\_to$ (Example 2.1). The database is represented by the set $\{\varepsilon_{IS_1}, \varepsilon_{IS_2}, \varepsilon_{IS_3}, \varepsilon_{IS_4}\}$, where

- $\varepsilon_{IS_1} = \{\varepsilon_{(name,\cdot)}, \varepsilon_{(age,\cdot)}, \varepsilon_{(gender,\cdot)}, \varepsilon_{(income,\cdot)}, \varepsilon_{(class,\cdot)}\}$:
  $SEM_{IS_1}(\varepsilon_{(name,\cdot)}) = \bigcup\{\{AS\} \times \{1\}, \{TJ\} \times \{2\}, \{AT\} \times \{3\}, \{SC\} \times \{4\}, \{ES\} \times \{5\}, \{JC\} \times \{6\}, \{MT\} \times \{7\}\}$, $SEM_{IS_1}(\varepsilon_{(age,\cdot)}) = \bigcup\{\{30\} \times \{1, 3, 4, 7\}, \{33\} \times \{2\}, \{26\} \times \{5\}, \{29\} \times \{6\}\}$, $SEM_{IS_1}(\varepsilon_{(gender,\cdot)}) = \bigcup\{\{m\} \times \{1, 6, 7\}, \{f\} \times \{2, 3, 4, 5\}\}$, $SEM_{IS_1}(\varepsilon_{(income,\cdot)}) = \bigcup\{\{1500\} \times \{1\}, \{1800\} \times \{3, 4, 7\}, \{2500\} \times \{2, 5\}, \{3000\} \times \{6\}\}$, $SEM_{IS_1}(\varepsilon_{(class,\cdot)}) = \bigcup\{\{yes\} \times \{1, 2, 4, 5, 6\}, \{no\} \times \{3, 7\}\}$.
- $\varepsilon_{IS_2} = \{\varepsilon_{(cust\_id,\cdot)}, \varepsilon_{(prod\_id,\cdot)}, \varepsilon_{(amount,\cdot)}, \varepsilon_{(date,\cdot)}\}$: $SEM_{IS_2}(\varepsilon_{(cust\_id,\cdot)}) = \bigcup\{\{1\} \times \{1, 2\}, \{2\} \times \{3, 4\}, \{4\} \times \{5, 6\}, \{5\} \times \{7\}, \{6\} \times \{8\}\}$, $SEM_{IS_2}(\varepsilon_{(prod\_id,\cdot)}) = \bigcup\{\{1\} \times \{1, 3\}, \{3\} \times \{2, 4\}, \{6\} \times \{5\}, \{2\} \times \{6\}, \{5\} \times \{7\}, \{4\} \times \{8\}\}$, $SEM_{IS_2}(\varepsilon_{(amount,\cdot)}) = \bigcup\{\{1\} \times \{1, 3, 4, 5, 6\}, \{2\} \times \{2, 7\}, \{3\} \times \{6\}\}$, $SEM_{IS_2}(\varepsilon_{(date,\cdot)}) = \bigcup\{\{24/6\} \times \{1, 2\}, \{25/06\} \times \{3\}, \{26/06\} \times \{4, 5, 6\}, \{27/06\} \times \{7, 8\}\}$.
- $\varepsilon_{IS_3} = \{\varepsilon_{(name,\cdot)}, \varepsilon_{(price,\cdot)}\}$:
  $SEM_{IS_3}(\varepsilon_{(name,\cdot)}) = \bigcup\{\{bread\} \times \{1\}, \{butter\} \times \{2\}, \{milk\} \times \{3\}, \{(tea\} \times \{4\}, \{coffee\} \times \{5\}, \{cigarettes\} \times \{6\}\}$, $SEM_{IS_3}(\varepsilon_{(price,\cdot)}) = \bigcup\{\{2.00\} \times \{1\}, \{3.50\} \times \{2\}, \{2.50\} \times \{3\}, \{5.00\} \times \{4\}, \{6.00\} \times \{5\}, \{6.5\} \times \{6\}\}$.
- $\varepsilon_{IS_4} = \{\varepsilon_{(cust\_id_1,\cdot)}, \varepsilon_{(cust\_id_1,\cdot)}\}$:
  $SEM_{IS_4}(\varepsilon_{(cust\_id_1,\cdot)}) = \bigcup\{\{5\} \times \{1\}, \{6\} \times \{2\}, \{3\} \times \{3\}\}$, $SEM_{IS_4}(\varepsilon_{(cust\_id_1,\cdot)}) = \bigcup\{\{1\} \times \{1\}, \{4\} \times \{2\}, \{7\} \times \{3\}\}$.

The connections among particular information systems are nor explicitly shown, however the granular representation includes information about how given systems can alternatively be joined. More formally.

Definition 8.5 (*Joinability of relations*) Let IS_i and IS_j be information systems. Relations $\varepsilon_{(a,\cdot)} \in L_{IS_i}^{\star}$ and $\varepsilon_{(a',\cdot)} \in L_{IS_j}^{\star}$ are joinable if and only if $\varepsilon_{(a,\cdot) \wedge (a',\cdot)} \in L_{IS_{(i,j)}}^{\star}$.

8.3.2   Relational Patterns Representation

To distinguish patterns constructed based on relations from those constructed based on formulas, we will call the former ε-relation based patterns (ε-patterns, in short). Let $D(\varepsilon_\alpha)$ and $D_a(\varepsilon_\alpha)$ denote, respectively, the domain of an ε-relation ε_α and the domain related to an attribute $a \in attr(\varepsilon_\alpha)$.[5]

Let $IS_{(m)} = \times (IS_1, IS_2, \ldots, IS_m)$ be a compound information system where each $IS_i = (U_i, A_i)$ is constructed based on a relation R_i.

8.3.2.1   Frequent Patterns

Frequent patterns are defined using ε-relation as follows.

Definition 8.6 (ε-*frequent pattern*)

1. An ε-pattern in $IS_{(m)}$ is a relation $\varepsilon_\alpha \in L_{IS_{(m)}}^{\star}$ such that $\alpha \in L_{IS_{(m)}}^{\star}$ is a pattern.[6]
2. The frequency of α is $freq_{IS_{(m)}}(\varepsilon_\alpha) = \frac{|SEM_{IS_{(m)}}(\varepsilon_\alpha)|}{|D(\varepsilon_\alpha)|}$.
3. The frequency of ε_α with respect to a domain $D_a(\varepsilon_\alpha)$ is $freq_{IS_{(m)}}^{\pi_a}(\varepsilon_\alpha) = \frac{|SEM_{IS_{(m)}}^{\pi_a}(\varepsilon_\alpha)|}{|D_a(\varepsilon_\alpha)|}$.

The above definition is a generalization of Definition 7.1. More formally.

Proposition 8.1 *Let $\varepsilon_\alpha \in L_{IS_{(m)}}^{\star}$ be an ε-pattern. If there exist $1 \leq i \leq m$ and $a \in attr(\varepsilon_\alpha)$ such that $U_i = D_a(\varepsilon_\alpha)$, then $freq_{IS_{(m)}}^{\pi_i}(\alpha) = freq_{IS_{(m)}}^{\pi_a}(\varepsilon_\alpha)$.*[7]

Definition 8.7 (*Frequency of ε-pattern under threshold*) Let $\varepsilon_\alpha \in L_{IS_{(m)}}^{\star}$ be an ε-pattern. The frequency of ε_α with respect to a domain $D_a(\varepsilon_\alpha)$ under a threshold $t \in [0, 1]$ imposed on a domain $D_{a'}(\varepsilon_\alpha)$ ($a \neq a'$) is defined as

$$freq_{IS_{(m)}}^{\pi_a}(\varepsilon_\alpha, t^{a',\#}) = \frac{|\{v \in D_a(\varepsilon_\alpha) : freq_{IS_{(m)}}^{\sigma_{a=v,\pi_{a'}}}(\varepsilon_\alpha)\#t\}|}{|D_a(\varepsilon_\alpha)|},$$

where $\# \in \{=, \neq, <, \leq, >, \geq\}$ and $freq_{IS_{(m)}}^{\sigma_{a=v,\pi_{a'}}}(\varepsilon_\alpha) = \frac{|SEM_{IS_{(m)}}^{\sigma_{a=v,\pi_{a'}}}(\varepsilon_\alpha)|}{|D_a(\varepsilon_\alpha)|}$.[8]

---

[5] $attr(\varepsilon_\alpha)$ denotes the set of all attributes used in an ε-relation ε_α.

[6] A pattern of the expanded language may include a descriptor of the form $(a, \cdot)$.

[7] Proofs of the propositions formulated in this section are simple and left to the reader.

[8] $\sigma_c(\bullet)$ is a selection under a condition c.

The above definition is a generalization of the pattern's frequency from Definition 8.6. More formally.

Proposition 8.2 *Let* $\varepsilon_\alpha \in L^*_{IS_{(m)}}$ *be any* ε-*pattern. The following holds*

$$\underset{a,a' \in attr(\varepsilon_\alpha), a \neq a'}{\forall} freq^{\pi_a}_{IS_{(m)}}(\varepsilon_\alpha) = freq^{\pi_a}_{IS_{(m)}}(\varepsilon_\alpha, t^{a',\geq}), \text{ where } t = 1/|D_a(\varepsilon_\alpha)|.$$

Based on one ε-pattern, we can acquire knowledge regarding more than one standard pattern.

Example 8.4 1. Let *IS* be the information system constructed based on relation *customer* from Example 2.1. Consider formula $\alpha_1 = (age, 30) \wedge (gender) \in L^*_{IS}$ and the relation ε_{α_1} with the schema $\varepsilon_{\alpha_1}(age, gender, id)$.
The semantics of ε_α is $SEM_{IS}(\varepsilon_{\alpha_1}) = \bigcup\{\{33\} \times \{female\} \times \{2\}, \{30\} \times \{female\} \times \{3, 4\}, \{26\} \times \{female\} \times \{5\}, \{29\} \times \{female\} \times \{6\}$.
We can acquire the following knowledge from ε_{α_1}.

   a. How often are customers females?
   $freq_{IS_1}(\varepsilon_{\alpha_1}) = \frac{|SEM_{IS_1}(\varepsilon_{\alpha_1})|}{|D(\varepsilon_{\alpha_1})|} = 4/7.$
   b. How often ages are associated with females?
   $freq^{\pi_{age}}_{IS_1}(\varepsilon_{\alpha_1}) = \frac{|SEM^{\pi_{age}}_{IS_1}(\varepsilon_{\alpha_1})|}{|D_{age}(\varepsilon_{\alpha_1})|} = 3/4.$

2. Let $IS_{(3)} = \times(IS_1, IS_2, IS_3)$ be the compound information system, where $IS_1, IS_2,$ and IS_3 are constructed based on relations $R_1 = customer,$ $R_2 = purchase,$ and $R_3 = product,$ respectively. Consider formula $\alpha_2 = (cust\_id, \cdot) \wedge (prod\_id, \cdot) \in L^*_{IS_{(3)}}$ and the relation ε_{α_2} with the schema $\varepsilon_{\alpha_2}(R_1.id, R_3.id, R_2.id)$.
The semantics of ε_{α_2} is $SEM_{IS_{(3)}}(\varepsilon_\alpha) = \bigcup\{\{1\} \times \{1\} \times \{1\}, \{1\} \times \{3\} \times \{2\}, \{2\} \times \{1\} \times \{3\}, \{2\} \times \{3\} \times \{4\}, \{4\} \times \{6\} \times \{5\}, \{4\} \times \{2\} \times \{6\}, \{5\} \times \{5\} \times \{7\}, \{6\} \times \{4\} \times \{8\}\}$.
We can acquire the following knowledge from ε_{α_2}.

   a. How often do customers purchase products?
   $freq^{\pi_{id_1}}_{IS_{(3)}}(\varepsilon_{\alpha_2}) = \frac{|SEM^{\pi_{id_1}}_{IS_{(3)}}(\varepsilon_{\alpha_2})|}{|U_1|} = \frac{|\{1,2,4,5,6\}|}{|\{1,...,7\}|} = 5/7.$[9]
   b. How often are products purchased by customers?
   $freq^{\pi_{id_3}}_{IS_{(3)}}(\varepsilon_{\alpha_2}) = \frac{|SEM^{\pi_{id_3}}_{IS_{(3)}}(\varepsilon_{\alpha_2})|}{|U_3|} = \frac{|\{1,...,6\}|}{|\{1,...,6\}|} = 1.$
   c. How often do customers purchase at least two products ($t_1 = 1/3$)?
   $freq^{\pi_{id_1}}_{IS_{(3)}}(\varepsilon_{\alpha_2}, t_1^{id_3,\geq}) = \frac{|\{x \in U_1 : freq^{\sigma_{id_1}=x,\pi_{id_3}}_{IS_{(3)}}(\varepsilon_{\alpha_2}) \geq t_1\}|}{|U_1|} = \frac{|\{1,2,4\}|}{|\{1,...,7\}|} = 3/7.$
   $freq^{\pi_{id_1}}_{IS_{(3)}}(\varepsilon_{\alpha_2}, t_1^{id_3,\geq}) = \frac{|\{x \in U_1 : freq^{\sigma_{id_1}=x,\pi_{id_3}}_{IS_{(3)}}(\varepsilon_{\alpha_2}) \geq t_1\}|}{|U_1|} = \frac{|\{1,2,4\}|}{|\{1,...,7\}|} = 3/7.$
   To compute the numerator we perform for each object from U_1 the following operations (shown for the first object): $SEM^{\sigma_{id_1}=1}_{IS_{(3)}}(\varepsilon_{\alpha_2}) = \bigcup\{\{1\} \times \{1\} \times$

---

[9] 1. For simplicity's sake we will write id_i for $R_i.id$. 2. The result means that 5 out of 7 customers purchase products.

$\{1\}, \{1\} \times \{3\} \times \{2\}\}, SEM_{IS_{(3)}}^{\sigma_{id_1=1},\pi_{id_3}}(\varepsilon_{\alpha_2}) = \{1,3\}$—the products purchased

by customer 1, hence $freq_{IS_{(3)}}^{\sigma_{id_1=1},\pi_{id_3}}(\varepsilon_{\alpha_2}) = \frac{|\{1,3\}|}{|\{1,...,6\}|} = 1/3$.

d. How often are products purchased by at least two customers ($t_2 = 2/7$)?

$$freq_{IS_{(3)}}^{\pi_{id_3}}(\varepsilon_{\alpha_2}, t_2^{id_1, \geq}) = \frac{|\{x \in U_3 : freq_{IS_{(3)}}^{\sigma_{id_3=x},\pi_{id_1}}(\varepsilon_{\alpha_2}) \geq t_2\}|}{|U_3|} = \frac{|\{1,3\}|}{|\{1,...,6\}|} = 1/3.$$

8.3.2.2 Association Rules

Association rules are defined using ε-relation as follows.

Definition 8.8 (ε-association rule)

1. An ε-association rule in $IS_{(m)}$ is an expression of the form $\varepsilon_{\alpha \to \beta} \in L_{IS_{(m)}}^\star$ such that $\alpha \to \beta \in L_{IS_{(m)}}^\star$ is an association rule in $IS_{(m)}$ and $\varepsilon_{\alpha \wedge \beta} \in L_{IS_{(m)}}^\star$ is an ε-pattern.
2. The frequency of $\varepsilon_{\alpha \to \beta}$ with respect to a domain of $D_a(\varepsilon_{\alpha \wedge \beta})$ is
$freq_{IS_{(m)}}^{\pi_a}(\varepsilon_{\alpha \to \beta}) = freq_{IS_{(m)}}^{\pi_a}(\varepsilon_{\alpha \wedge \beta})$.
3. The confidence of $\varepsilon_{\alpha \to \beta}$ with respect to a domain of $D_a(\varepsilon_\alpha)$ is
$conf_{IS_{(m)}}^{\pi_a}(\varepsilon_{\alpha \to \beta}) = \frac{freq_{IS_{(m)}}^{\pi_a}(\varepsilon_{\alpha \wedge \beta})}{freq_{IS_{(m)}}^{\pi_a}(\varepsilon_\alpha)}$. [10]

Definition 8.9 (*Confidence of ε-association rule under threshold*) Let $\varepsilon_{\alpha \to \beta} \in L_{IS_{(m)}}^\star$ be an ε-association rule. The confidence of $\varepsilon_{\alpha \to \beta}$ with respect to a domain $D_a(\varepsilon_\alpha)$ under a threshold $t \in [0, 1]$ imposed on a domain $D_{a'}(\varepsilon_\alpha)$ ($a \neq a'$) is defined as

$$conf_{IS_{(m)}}^{\pi_a}(\varepsilon_{\alpha \to \beta}, t^{a',\#}) = \frac{|\{v \in D_a(\varepsilon_\alpha) : freq_{IS_{(m)}}^{\sigma_{a=v},\pi_{a'}}(\varepsilon_{\alpha \wedge \beta}) \# t\}|}{|SEM_{IS_{(m)}}^{\pi_a}(\varepsilon_\alpha)|}.$$

Definition 8.10 (*Confidence of ε-association rule under double threshold*) Let $\varepsilon_{\alpha \to \beta} \in L_{IS_{(m)}}^\star$ be an ε-association rule. The confidence of $\varepsilon_{\alpha \to \beta}$ with respect to the a domain $D_a(\varepsilon_\alpha)$ under a threshold $t \in [0, 1]$ imposed doubly on a domain $D_{a'}(\varepsilon_\alpha)$ ($a \neq a'$) is defined as

$$conf_{IS_{(m)}}^{\pi_a}(\varepsilon_{\alpha \to \beta}, 2 * t^{\#,a'}) = \frac{freq_{IS_{(m)}}^{\pi_a}(\varepsilon_{\alpha \wedge \beta}, t^{a',\#})}{freq_{IS_{(m)}}^{\pi_a}(\varepsilon_\alpha, t^{a',\#})} =$$

$$\frac{|\{v \in D_a(\varepsilon_\alpha) : freq_{IS_{(m)}}^{\sigma_{a=v},\pi_{a'}}(\varepsilon_{\alpha \wedge \beta}) \# t\}|}{|\{v \in D_a(\varepsilon_\alpha) : freq_{IS_{(m)}}^{\sigma_{a=v},\pi_{a'}}(\varepsilon_\alpha) \# t\}|}.$$

Analogously to frequent patterns, Definition 8.8 is a generalization of Definition 7.2, whereas Definitions 8.9 and 8.10 generalize the rule's confidence from Definition 8.8.

---

[10]Unlike for patterns, the frequency and confidence of an association rule $\varepsilon_{\alpha \to \beta}$ with respect to the domain $D(\varepsilon_{\alpha \to \beta})$ are not defined, since $D(\varepsilon_\alpha)$ differs from $D(\varepsilon_\beta)$.

Example 8.5 Let $IS_{(3)}$ be the compound information system defined as in Example 8.4. Consider formulas $\alpha = (cust\_id, \cdot) \wedge (prod\_id, \cdot)$, $\beta = (cust\_id, \cdot) \wedge (prod\_id, \cdot) \wedge (amount, 1) \in L^{\star}_{IS_{(3)}}$ and the relations with the schemas $\varepsilon_{\alpha}(R_1.id, R_3.id, R_2.id)$, $\varepsilon_{\beta}(R_1.id, R_3.id, amount, R_2.id)$.

The semantics of ε_{α} and ε_{β} are $SEM_{IS_{(3)}}(\varepsilon_{\alpha}) = \bigcup\{\{1\}\times\{1\}\times\{1\}, \{1\}\times\{3\}\times\{2\}, \{2\}\times\{1\}\times\{3\}, \{2\}\times\{3\}\times\{4\}, \{4\}\times\{6\}\times\{5\}, \{4\}\times\{2\}\times\{6\}, \{5\}\times\{5\}\times\{7\}, \{6\}\times\{4\}\times\{8\}\}$ and $SEM_{IS_{(3)}}(\varepsilon_{\beta}) = \bigcup\{\{1\}\times\{1\}\times\{1\}\times\{1\}, \{2\}\times\{1\}\times\{1\}\times\{3\}, \{2\}\times\{3\}\times\{1\}\times\{4\}, \{4\}\times\{6\}\times\{1\}\times\{5\}, \{6\}\times\{4\}\times\{1\}\times\{8\}\}$.

We can acquire the following knowledge from $\varepsilon_{\alpha\rightarrow\beta}$.

1. How are customers who purchase products likely to purchase the products in quantities of one piece?

$$conf^{\pi_{id_1}}_{IS_{(3)}}(\varepsilon_{\alpha\rightarrow\beta}) = \frac{|SEM^{\pi_{id_1}}_{IS_{(3)}}(\varepsilon_{\beta})|}{|SEM^{\pi_{id_1}}_{IS_{(3)}}(\varepsilon_{\alpha})|} = \frac{|\{1,2,4,6\}|}{|\{1,2,4,5,6\}|} = 4/5.^{11}$$

2. How are products being purchased likely to be purchased in quantities of one piece by customers?

$$conf^{\pi_{id_3}}_{IS_{(3)}}(\varepsilon_{\alpha\rightarrow\beta}) = \frac{|SEM^{\pi_{id_3}}_{IS_{(3)}}(\varepsilon_{\beta})|}{|SEM^{\pi_{id_3}}_{IS_{(3)}}(\varepsilon_{\alpha})|} = \frac{|\{1,3,4,6\}|}{|\{1,...,6\}|} = 2/3.$$

3. How are customers who purchase products likely to purchase at least two of the products ($t_1 = 1/3$) in quantities of one piece?

$$conf^{\pi_{id_1}}_{IS_{(3)}}(\varepsilon_{\alpha\rightarrow\beta}, t_1^{id_3,\geq}) = \frac{|\{x\in U_1 : freq^{\sigma_{id_1}=x \cdot \pi_{id_3}}_{IS_{(3)}}(\varepsilon_{\alpha\wedge\beta})\geq t_1\}|}{|SEM^{\pi_{id_1}}_{IS_{(3)}}(\varepsilon_{\alpha})|} = \frac{|\{2\}|}{|\{1,2,4,5,6\}|} = 1/4.$$

4. How are products being purchased likely to be purchased by at least two customers ($t_2 = 2/7$) in quantities of one piece?

$$conf^{\pi_{id_3}}_{IS_3}(\varepsilon_{\alpha\rightarrow\beta}, t_2^{id_1,\geq}) = \frac{|\{x\in U_3 : freq^{\sigma_{id_3}=x \cdot \pi_{id_1}}_{IS_{(3)}}(\varepsilon_{\alpha\wedge\beta})\geq t_2\}|}{|SEM^{\pi_{R_3.id}}_{IS_{(3)}}(\varepsilon_{\alpha})|} = \frac{|\{1\}|}{|\{1,...,6\}|} = 1/6.$$

5. How are customers who purchase at least two products ($t_3 = 1/3$) likely to purchase all these products in quantities of one piece?

$$conf^{\pi_{id_1}}_{IS_{(3)}}(\varepsilon_{\alpha\rightarrow\beta}, 2 * t_3^{id_3,\geq}) = \frac{|\{x\in U_1 : freq^{\sigma_{id_1}=x}_{IS_{(3)}}(\varepsilon_{\alpha\wedge\beta})\geq t_3\}|}{|\{x\in U_1 : freq^{\sigma_{id_1}=x \cdot \pi_{id_3}}_{IS_{(3)}}(\varepsilon_{\alpha})\geq t_3\}|} = \frac{|\{2\}|}{|\{1,2,4\}|} = 1/3.$$

6. How are products being purchased by at least two customers ($t_4 = 2/7$) likely to be purchased by all these customers in quantities of one piece?

$$conf^{\pi_{id_3}}_{IS_{(3)}}(\varepsilon_{\alpha\rightarrow\beta}, 2 * t_4^{id_1,\geq}) = \frac{|\{x\in U_3 : freq^{\sigma_{id_3}=x \cdot \pi_{id_1}}_{IS_1}(\varepsilon_{\alpha\wedge\beta})\geq t_4\}|}{|\{x\in U_3 : freq^{\sigma_{id_3}=x \cdot \pi_{id_1}}_{IS_1}(\varepsilon_{\alpha})\geq t_4\}|} = \frac{|\{1\}|}{|\{1,3\}|} = 1/2.$$

8.3.2.3 Classification Rules

Classification rules are defined using ε-relation as follows.

Definition 8.11 (*ε-classification rule*)

1. An ε-classification rule in $IS_{(m)}$ is an ε-association rule $\varepsilon_{\alpha\rightarrow\beta} \in L^{\star}_{IS_{(m)}}$ such that β is the decision descriptor.

---

[11] The result means that 4 out of 5 customers who purchase products purchase them in quantities of one piece.

2. The accuracy of $\varepsilon_{\alpha \to \beta}$ with respect to a domain of $D_a(\varepsilon_\alpha)$ is $acc_{IS_{(m)}}^{\pi_a}(\varepsilon_{\alpha \to \beta}) = conf_{IS_{(m)}}^{\pi_a}(\varepsilon_{\alpha \to \beta})$.

3. The coverage of $\varepsilon_{\alpha \to \beta}$ with respect to a domain of $D_a(\varepsilon_\alpha)$ such that $D_a(\varepsilon_\beta)$ exists is $cov_{IS_{(m)}}^{\pi_a}(\varepsilon_{\alpha \to \beta}) = \frac{freq_{IS_{(m)}}^{\pi_a}(\varepsilon_{\alpha \wedge \beta})}{freq_{IS_{(m)}}^{\pi_a}(\varepsilon_\beta)}$.

Since a classification rule is a special case of an association one, then its accuracy under a threshold is defied as the association rule's confidence (see Definitions 8.9 and 8.10).

Example 8.6 Let $IS_{(3)}$ be the compound information system defined as in Example 8.4. Consider formulas $\alpha = (cust\_id, \cdot) \wedge (prod\_id, \cdot)$, $\beta = (class, yes) \in L_{IS_{(3)}}$ and the relations $\varepsilon_\alpha, \varepsilon_\beta$ with the schemas $\varepsilon_\alpha(R_1.id, R_3.id, R_2.id)$, $\varepsilon_\beta(class, R_i.id)$.
The semantics of ε_α and ε_β are $SEM_{IS_{(3)}}(\varepsilon_\alpha) = \bigcup \{\{1\} \times \{1\} \times \{1\}, \{1\} \times \{3\} \times \{2\}, \{2\} \times \{1\} \times \{3\}, \{2\} \times \{3\} \times \{4\}, \{4\} \times \{6\} \times \{5\}, \{4\} \times \{2\} \times \{6\}, \{5\} \times \{5\} \times \{7\}, \{6\} \times \{4\} \times \{8\}\}$ and $SEM_{IS_{(3)}}(\varepsilon_\beta) = \bigcup \{\{yes\} \times \{1, 2, 4, 5, 6\}\}$.
We can acquire the following knowledge from $\varepsilon_{\alpha \to \beta}$.

1. How are customers who purchase products likely to be considered as good customers?
$$acc_{IS_{(3)}}^{\pi_{id_1}}(\varepsilon_{\alpha \to \beta}) = \frac{|SEM_{IS_{(3)}}^{\pi_{id_1}}(\varepsilon_{\alpha \wedge \beta}))|}{|SEM_{IS_{(3)}}^{\pi_{id_1}}(\varepsilon_\alpha)|} = \frac{|\{1,2,4,5\}|}{|\{1,2,4,5,6\}|} = 4/5.^{12}$$

2. How are customers who purchase at least two products ($t = 1/3$) likely to be considered as good customers?
$$acc_{IS_{(3)}}^{\pi_{id_1}}(\varepsilon_{\alpha \to \beta}, t^{2 \cdot a, \geq}) = \frac{|\{x \in U_1 : freq_{IS_{(3)}}^{\sigma_{id_1 = x}, \pi_{id_3}}(\varepsilon_{\alpha \wedge \beta}) \geq t\}|}{|SEM_{IS_{(3)}}^{\pi_{id_1}}(\varepsilon_\alpha)|} = \frac{|\{1,2,4\}|}{|\{1,2,4,5,6\}|} = 3/5.$$

8.4 The Approach's Complexity

This section evaluates the cost of the construction of representations of relational data and patterns using introduced relations. It also compares the approach introduced in this chapter with a standard one in terms of complexity. The latter is understood as an approach where database tables are mined directly, i.e. the data is not transformed into an alternative representation (except for using typical data mining transformation techniques such as e.g. discretization).

Table 8.1 includes the cost of database transformation and the cost of checking the satisfiability of formulas (patterns) during their construction. As it can be seen, a high cost of the database transformation is offset by a lower cost of pattern genera-

---

[12]The result means that 4 out of 5 customers who purchase products are considered as good customers.

Table 8.1 Complexity of operations for the granular (GA) and standard (SA) approaches. Operations being compared are (the labels with primes denote equivalent operations for the standard approach): op_1—forming the set $\{\varepsilon_{IS_i} : 1 \le i \le m\}$ that represents $IS_{(m)}$; op_2—checking a formula $(a, v) \in L_{IS}$; op_3—checking a formula $(a, v) \wedge (a', v') \in L_{IS}$; op_4—checking a formula $(a, \cdot) \in L_{IS}$; op_5—checking a formula $(a, \cdot) \wedge (a', \cdot) \in L_{IS}$; op_6—checking a formula $(a, \cdot) \wedge (a', v') \in L_{IS_{(i,j)}}$, where $(a, \cdot) \in L_{IS_i}$ and $(a', v') \in L_{IS_j}$; op_7—checking a formula $(a, \cdot) \wedge (a', \cdot) \in L_{IS_{(i,j)}}$, where $(a, \cdot) \in L_{IS_i}$ and $(a', \cdot) \in L_{IS_j}$

GA		SA	
op_1	$O((n^{max})^2)$	$op_{1'}$	$O(1)$
op_2	$O(1)$	$op_{2'}$	$O(n)$
op_3	$O(n)$	$op_{3'}$	$O(n)$
op_4	$O(1)$	$op_{4'}$	$O(n)$
op_5	$O(n)$	$op_{5'}$	$O(n)$
op_6	$O(n^{max})$	$op_{6'}$	$O((n^{max})^2)$
op_7	$O((n^{max})^2)$	$op_{7'}$	$O((n^{max})^2)$

tion. Namely, the database is transformed only once regardless of the task (frequent patterns/association discovery or classification), but patterns can be generated repeatedly. Therefore, granular database representation make it possible to speed up the generation of patterns.

8.4.1 The Granular Approach's Complexity

The further part of this section shows the details of the computational costs of all operations. Firstly, the way of the construction of a granular representation of relational data is evaluated.

Consider $IS = (U, A)$ including only descriptive attributes except for the identifier, i.e. $A_{des} = A \backslash \{id\}$. Let $n = |U|$.

It is assumed that the cost of forming the condition (a, v) where attribute $a \in A$ and value $v \in V_a$ are given is 1.

- The cost of forming a relation $\varepsilon_{(a,\cdot)} \in L_{IS}^{\star}$, where $a \in A_{des}$,[13] is

$$T_1^{des}(n) < Cn = O(n),$$

where $C = |V_a| - 1$. For each $o \in U$ we scan the list of values used so far to check if $v = a(o)$ is a new value (the list cardinality is less than or equal to $|V_a|$). It is assumed that $|V_a|$ is relatively small by nature or thanks to a discretization.
- The cost of forming the tuple ε_{IS} that represents IS is

---

[13]The cost of forming $\varepsilon_{(id,\cdot)}$ is n because $SEM_{IS}(\varepsilon_{(id,\cdot)}) = \{(id(x), id(x)) : x \in U\}$.

$$T_2^{des}(n) = |A|T_1(n) = O(n).$$

Given a compound information system $IS_{(m)} = \times(IS_1, IS_2, \ldots, IS_m)$. Each $IS_i = (U_i, A_i)$ $(1 \leq i \leq m)$ is constructed based on a database relation R_i. Let $n_i = |U_i|$. Consider now an information system IS_i including also key attributes other the identifier, i.e. there exists $a \in A_{key}\backslash\{id\}$ such that $a = R_j.id$ and $j \neq i$. We obtain that $V_a \subseteq U_j$.

- The cost of forming a relation $\varepsilon_{(a,\cdot)} \in L_{IS_i}^\star$, where $a \in A_{key}\backslash\{id\}$, is

$$T_1^{key}(n_1, n_2, \ldots, n_m) < |V_a|n_i < n_j n_i < n^{max}n_i = O(n^{max}n_i),$$

where $n^{max} = max\{n_i : 1 \leq i \leq m\}$.
- The cost of forming the tuple ε_{IS_i} that represents IS_i is

$$T_2^{key}(n_1, n_2, \ldots, n_m) = C_1 T_1^{des}(n_i) + C_2 T_1^{key}(n_1, n_2, \ldots, n_m) = O(n^{max}n_i),$$

where $C_1 = |A_{des}| + 1$ and $C_2 = |A_{key}\backslash\{id\}|$.
- (op_1) The cost of forming the set $\{\varepsilon_{IS_i} : 1 \leq i \leq m\}$ that represents $IS_{(m)}$ is

$$T_3(n_1, n_2, \ldots, n_m) = \sum_{i=i_1}^{i_{m_1}} T_2^{des}(n_{i_1}) + \sum_{j=j_1}^{j_{m_1}} T_2^{key}(n_1, n_2, \ldots, n_m) = m_1 O(n^{max}) + m_2 O((n^{max})^2) = O((n^{max})^2),$$

where $i_1, i_2, \ldots, i_{m_1}$ are labels of information systems including only descriptive attributes except for the identifier, $j_1, j_2, \ldots, j_{m_1}$, are labels of the remaining information systems, and $m_1 + m_2 = m$.

The way of checking the satisfiability of formulas during their construction is evaluated as follows.
It is assumed that the cost of checking the condition (a, v) for an object $o \in U$ where attribute $a \in A$ and value $v \in V_a$ are given is 1.
Given an information system $IS = (U, A)$.

- (op_2) The cost of checking a formula $(a, v) \in L_{IS}^{\star}{}^{14}$ is

$$T_4(n) \leq |V_a| = O(1).$$

We only need to scan the semantics of $\varepsilon_{(a,\cdot)}$ to find the set corresponding to value $v \in V_a$.
- (op_3) The cost of checking a formula $(a, v) \wedge (a', v') \in L_{IS}^\star$ is

$$T_5(n) \leq |V_a| + |V_{a'}| + 2n = O(n).$$

---

[14]Formulas are constructed over descriptive attributes only. Key attributes are used in a compound information system to join particular information systems.

We scan the semantics of $\varepsilon_{(a,\cdot)}$ (at most $|V_a|$ operations) and $\varepsilon_{(a',\cdot)}$ (at most $|V_{a'}|$ operations) to find the sets corresponding to v and v' and then compute the intersection of them (at most $2n$ operations[15]).

- (op_4) The cost of checking a formula $(a, \cdot) \in L_{IS}^{\star}$ is

$$T_6(n) = |V_a| = O(1).$$

We only need to scan the semantics of $\varepsilon_{(a,\cdot)}$ to get the set corresponding to each value $v \in V_a$.

- The cost of checking a formula $(a, \cdot) \wedge (a', v') \in L_{IS}^{\star}$ is

$$T_7(n) \leq |V_{a'}| + 2n|V_a| = O(n).$$

We scan the semantics of $\varepsilon_{(a',\cdot)}$ (at most $|V_a|$ operations) to find the set corresponding to v' and then compute the intersection of each pair of sets corresponding to v' and $v \in V_a$ (at most $2n|V_a|$).

- (op_5) The cost of checking a formula $(a, \cdot) \wedge (a', \cdot) \in L_{IS}^{\star}$ is

$$T_8(n) = 2n|V_a||V_{a'}| = O(n).$$

For each pair $(v, v') \in V_a \times V_{a'}$ we compute the intersection of the sets corresponding to v and v'.

Given an information system $IS_{(i,j)} = \times (IS_i, IS_j)$.

- The cost of checking a formula $(a, v) \wedge (a', v') \in L_{IS_{(i,j)}}$, where $(a, v) \in L_{IS_i}^{\star}$ and $(a', v') \in L_{IS_j}$ is

$$T_9(n_i, n_j) = max\{n_i, n_j\} = O(n^{max}),$$

where $n^{max} = max\{n_i, n_j\}$. If $v' \in SEM_{IS_i}(a, v)$, we have $T_9(n_i, n_j) = n_i$. The sets corresponding to all $v \in V_a$ form a partition of U_i. We need to scan the semantics of $\varepsilon_{(a,\cdot)}$ to find $v \in V_a$ such that v' belongs to the set corresponding to v. Therefore, this is equivalent to scanning U_i. If $v \in SEM_{IS_j}(a', v')$ we analogously obtain $T_9(n_i, n_j) = n_j$.

- (op_6) The cost of checking a formula $(a, \cdot) \wedge (a', v') \in L_{IS_{(i,j)}}^{\star}$, where $(a, \cdot) \in L_{IS_i}^{\star}$ and $(a', v') \in L_{IS_j}^{\star}$ is

$$T_{10}(n_i, n_j) = max\{n_i, 2n_j\} = O(n^{max}).$$

If $v' \in SEM_{IS_i}(a, v)$, we obtain $T_{10}(n_i, n_j) = n_i$ analogously to the previous point. If $v \in SEM_{IS_j}(a', v')$ we have $T_{10}(n_i, n_j) = 2n_j$. We scan V_a (at most n_i operations since $V_a \subseteq U_j$) and at the same time the set corresponding to v' to check if a given

---

[15]The sets are assumed to be ordered. This operation does not increase the asymptotic complexity of the database transformation.

v from V_a belongs to the set (at most n_i operations[16]).
We obtain an analogous result for $(a, v) \wedge (a', \cdot) \in L^*_{IS_{(i,j)}}$.

- (op_7) The cost of checking a formula $(a, \cdot) \wedge (a', \cdot) \in L^*_{IS_{(i,j)}}$, where $(a, \cdot) \in L^*_{IS_i}$ and $(a', \cdot) \in L^*_{IS_j}$ is

$$T_{11}(n_i, n_j) \leq max\{|V_a|n_in_j, |V_{a'}|n_in_j\} = O((n^{max})^2).$$

If $v' \in SEM_{IS_i}(a, v)$, we have $T_{10}(n_i, n_j) = |V_a||V_{a'}|n_i \leq |V_a|n_in_j$. For each pair $(v, v') \in V_a \times V_{a'}$ ($|V_a||V_{a'}|$ operations) we scan the set corresponding to v to check if v' belongs to this set (at most n_i operations). Since $V_{a'} \subseteq U_i$, then we obtain $|V_a||V_{a'}|n_i \leq |V_a|n_j$. If $v \in SEM_{IS_j}(a', v')$ we analogously obtain $T_{10}(n_i, n_j) \leq |V_{a'}|n_in_j$.

8.4.2   The Standard Approach's Complexity

The following compares the introduced approach with a standard one in terms of complexity. In a standard approach relational data is mined as is, i.e. no alternative representation is generated. Therefore, the cost of the database transformation is $O(1)$ ($op_{1'}$).

Consider a relation R based on which $IS = (U, A)$ is formed. We have that the cardinality of R is n.

- $(op_{2'})$ The cost of checking a formula (a, v) is

$$T_{4'}(n) = n = O(n).$$

We need to scan all objects from R.
- $(op_{3'})$ The cost of checking a formula $(a, v) \wedge (a', v')$ is

$$T_{5'}(n) \leq 2n = O(n).$$

For all objects that satisfy (a, v) (at most n objects) we need to scan R to check if they satisfy (a', v') (n operations).
- $(op_{4'})$ The cost of checking a formula $\bigvee\limits_{v \in V_a} (a, v)$ (an equivalent of $(a, \cdot) \in L_{IS}$) is

$$T_{6'}(n) = |V_a|n = O(n).$$

When searching for the best descriptor constructed based on an attribute a, in fact, we examine each formula (a, v), where $v \in V_a$ (a classification rules generation case).

---

[16]We assume that the sets are ordered.

- $(op_{5'})$ The cost of checking a formula $\bigvee_{v \in V_a} \bigvee_{v \in V_a} (a, v) \wedge (a', v')$ (an equivalent of $(a, \cdot) \wedge (a', \cdot) \in L_{IS}$) is

$$T_{8'}(n) = |V_a||V_{a'}|T_{5'}(n) = O(n).$$

The formula $\bigvee_{v \in V_a} \bigvee_{v \in V_a} (a, v) \wedge (a', v)'$ corresponds to the set of all formulas constructed over both the attributes a and a' (a frequent patterns generation case).

Consider relations R_i and R_j based on which $IS_i = (U_i, A_i)$ and $IS_j = (U_j, A_j)$ are formed. We have that cardinalities of R_i and R_j are n_i and n_j, respectively.

- $(op_{6'})$ The cost of checking a formula $(a, v) \wedge (R_j.a' = R_i.id)$ (an equivalent of $(a, v) \wedge (a', \cdot) \in L_{IS(i,j)}$) is

$$T_{10'}(n_i, n_j) \leq n_i + n_i n_j = O((n^{max})^2).$$

For each object that satisfies (a, v) (at most n_i objects) we scan R_j to check if the object also satisfies $(R_j.a' = R_i.id)$ (n_j operations).[17]
- $(op_{7'})$ The cost of checking a formula $\bigvee_{v \in V_a} (a, v) \wedge (R_j.a' = R_i.id)$ (an equivalent of $(a, \cdot) \wedge (a', \cdot) \in L_{IS(i,j)}$) is

$$T_{11'}(n_i, n_j) \leq |V_a|T_{10'}(n_i, n_j) = O((n^{max})^2).$$

8.5 Conclusions

This chapter has expended description languages defined for relational information granules. The expansion includes formula-based relations designed for representing relational databases and patterns to be discovered. The main advantages of the approach can be summarized as follows.

1. The cost of generation of relational patterns can be decreased compared with that when the patterns are generated directly from the database. In fact, relations that represent the database consist of atomic formulas to be used for pattern construction.
2. Richer knowledge can be discovered from relational data when generating patterns using relations. The patterns exploit information which can be acquired from a standard relational database by applying additional computations such as aggregation.

---

[17]Relational data in the form it is provided is not, in general, ordered.

Chapter 9
Compound Approximation Spaces

9.1 Introduction

Constructing a rough set model for processing data stored in a relational structure is not a trivial task. A relational database considered in the context of data mining tasks (e.g. classification) has a specified table (target table) that includes objects to be analyzed and it can be treated as the counterpart of the single table database. The remaining relational database tables (background tables) include additional data that is directly or indirectly associated with the target table. For that reason, a lot of, or even most, essential information about target objects can be hidden in the background tables.

The crucial problem when applying rough sets to relational data is, therefore, to construct an approximation space. Such a space should include essential information about target objects, background objects, as well as relationships among them.

The goal of this chapter is to introduce a framework for processing relational data using rough set tools [43]. The underlying idea is to use the benefit of rough set theory to deal with uncertainty in relational data. Such an uncertainty may concern not only objects of a given database table but also the relationship of objects from different tables.

The chapter develops compound approximation spaces and their constrained versions that are constructed over relational data. The universe in a compound approximation space is the Cartesian product of the universes of particular approximation spaces (each corresponding to one database table). The universe in a constrained compound approximation space is limited according to possible connections between database tables. The lower and upper approximations are defined for (constrained) compound concepts that are subsets of the (constrained) compound universe.

The remaining of the chapter is organized as follows. Section 9.2 introduces compound approximation spaces. Section 9.3 studies acquisition of knowledge from approximations of compound. Section 9.4 evaluates the complexity of the approach. Section 9.5 provides concluding remarks.

© Springer International Publishing AG 2017
P. Hońko, *Granular-Relational Data Mining*, Studies in Computational
Intelligence 702, DOI 10.1007/978-3-319-52751-2_9

9.2 Compound Approximation Spaces and Their Constrained Versions

This section introduces compound approximation spaces that are defined for compound information systems introduced in the previous section.

Firstly, the notion of approximation space is slightly redefined.

Definition 9.1 (*Approximation space AS_ω*) An approximation space AS_ω for an information system $IS = (U, A)$ is defined by

$$AS_\omega = (U, I_\omega, v_\omega) \tag{9.1}$$

where $\omega = (\#, \$)$, $I_\omega = I_\#$, $v_\omega = v_\$$.

9.2.1 Compound Approximation Spaces

A compound approximation space corresponding to two database tables is defined as follows.

Definition 9.2 (*Compound approximation space $AS_{\omega_{(i,j)}}$*) Let $AS_{\omega_i} = (U_i, I_{\omega_i}, v_{\omega_i})$ and $AS_{\omega_j} = (U_j, I_{\omega_j}, v_{\omega_j})$ ($i \neq j$) be approximation spaces for information systems $IS_i = (U_i, A_i)$ and $IS_j = (U_j, A_j)$, respectively. A compound approximation space (CAS for short) $AS_{\omega_{(i,j)}}$ for a compound information system $IS_{(i,j)} = \times(IS_i, IS_j)$ is defined by

$$AS_{\omega_{(i,j)}} = \times(AS_{\omega_i}, AS_{\omega_j}) = (U_{\omega_{(i,j)}}, I_{\omega_{(i,j)}}, v_{\omega_{(i,j)}}) \tag{9.2}$$

where

- $U_{\omega_{(i,j)}} = U_i \times U_j$,
- $\displaystyle\forall_{(x_1,x_2) \in U_{\omega_{(i,j)}}} \quad I_{\omega_{(i,j)}}((x_1, x_2)) = I_{\omega_i}(x_1) \times I_{\omega_j}(x_2)$,
- $\displaystyle\forall_{X_1, Y_1 \in U_i, X_2, Y_2 \in U_j} \quad v_{\omega_{(i,j)}}(X_1 \times X_2, Y_1 \times Y_2) = v_{\omega_i}(X_1, Y_1) \cdot v_{\omega_j}(X_2, Y_2)$.

For each of the universes U_i and U_j a different uncertainty function and rough inclusion function can be used. If we apply the standard rough inclusion to both the universes we obtain the following relationship.

Proposition 9.1 [1] *Let $AS_{\omega_{(i,j)}}$ be a CAS and $v_i = v_j = v_{SRI}$. The following holds*
$$\forall_{X_1, Y_1 \subseteq U_i, X_2, Y_2 \subseteq U_j} \quad v_{\omega_{(i,j)}}(X_1 \times X_2, Y_1 \times Y_2) = v_{SRI}(X_1 \times X_2, Y_1 \times Y_2).$$

Approximations of a set in a compound approximation space are defined as follows.

---

[1] Proofs of the propositions formulated in this chapter can be found in [43].

Definition 9.3 (*Approximations of a set in* $AS_{\omega_{(i,j)}}$) Let $AS_{\omega_{(i,j)}} = (U_{\omega_{(i,j)}}, I_{\omega_{(i,j)}}, \nu_{\omega_{(i,j)}})$ be a CAS and $X_1 \subseteq U_i, X_2 \subseteq U_j$. The lower and upper approximations of the set $X_1 \times X_2$ in $AS_{\omega_{(i,j)}}$ are defined, respectively, by

$$LOW(AS_{\omega_{(i,j)}}, X_1 \times X_2) = \{(x_1, x_2) \in U_{\omega_{(i,j)}} : \nu_{\omega_{(i,j)}}(I_{\omega_{(i,j)}}((x_1, x_2)), X_1 \times X_2) = 1\},$$

$$UPP(AS_{\omega_{(i,j)}}, X_1 \times X_2) = \{(x_1, x_2) \in U_{\omega_{(i,j)}} : \nu_{\omega_{(i,j)}}(I_{\omega_{(i,j)}}((x_1, x_2)), X_1 \times X_2) > 0\}.$$

The lower and upper approximations possess the following properties.

Proposition 9.2 *Let* $AS_{\omega_{(i,j)}} = \times(AS_{\omega_i}, AS_{\omega_j})$ *be a CAS. The following hold*
$$\underset{X_1 \subseteq U_i, X_2 \subseteq U_j}{\forall}$$

1. $LOW(AS_{\omega_i}, X_1) \times U_j = LOW(AS_{\omega_{(i,j)}}, X_1 \times U_j)$,
2. $UPP(AS_{\omega_i}, X_1) \times U_j = UPP(AS_{\omega_{(i,j)}}, X_1 \times U_j)$,
3. $LOW(AS_{\omega_j}, X_2) \neq \emptyset \Rightarrow \pi_i(LOW(AS_{\omega_{(i,j)}}, X_1 \times X_2)) = LOW(AS_{\omega_i}, X_1)$.
4. $UPP(AS_{\omega_j}, X_2) \neq \emptyset \Rightarrow \pi_i(UPP(AS_{\omega_{(i,j)}}, X_1 \times X_2)) = UPP(AS_{\omega_i}, X_1)$.
5. $LOW(AS_{\omega_i}, X_1) \times LOW(AS_{\omega_j}, X_2) = LOW(AS_{\omega_{(i,j)}}, X_1 \times X_2)$,
6. $UPP(AS_{\omega_i}, X_1) \times UPP(AS_{\omega_j}, X_2) = UPP(AS_{\omega_{(i,j)}}, X_1 \times X_2)$.

For AS_{ω_j} one can formulate equalities analogous to 1–4.

Example 9.1 Consider approximation spaces $AS_{\omega_{(1,2)}} = \times(AS_{\omega_1}, AS_{\omega_2})$, where AS_{ω_1} and AS_{ω_2} are constructed respectively based on relations *customer* and *purchase* from Example 2.1. Let $I_{\omega_1} = I_{\{age,income\},\varepsilon_1}$, $I_{\omega_2} = I_{\{amount,date\},\varepsilon_2}$, $\varepsilon_1 = (\varepsilon_{age}, \varepsilon_{income}) = (2, 300)$, $\varepsilon_2 = (\varepsilon_{amount}, \varepsilon_{date}) = (1, 1)$, $\nu_{\omega_1} = \nu_{SRI}$ and $\nu_{\omega_2} = \nu_{u,l}$ where $u = 0.25$ and $l = 0.75$.[2] Let $X_1 = \{1, 2, 4, 7\} \subset U_1$, $X_2 = \{2, 6, 7\} \subset U_2$, and $X_1 \times X_2$ be the sets to be approximated.
The table below shows the similarity classes and their rough inclusion degrees in the respective sets.

$x_1 \in U_1$	$I_{\omega_1}(x_1)$	$\nu_{\omega_1}(I_{\omega_1}(x_1), X_1)$	$x_2 \in U_2$	$I_{\omega_2}(x_2)$	$\nu_{\omega_2}(I_{\omega_2}(x_2), X_2)$
1	$\{1\}$	1	1	$\{1, 2, 3\}$	0.33
2	$\{2\}$	1	2	$\{1, 2, 3\}$	0.33
3	$\{3, 4\}$	0.5	3	$\{1, \ldots, 5\}$	0.2
4	$\{3, 4\}$	0.5	4	$\{3, 4, 5, 8\}$	0
5	$\{5\}$	1	5	$\{3, 4, 5, 8\}$	0
6	$\{6\}$	1	6	$\{6, 7\}$	1
7	$\{7\}$	0	7	$\{6, 7\}$	1
–	–	–	8	$\{4, 5, 8\}$	0

We obtain the following approximations $LOW(AS_{\omega_1}, X_1) = \{1, 2, 5, 6\}$, $UPP(AS_{\omega_1}, X_1) = \{1, \ldots, 6\}$, $LOW(AS_{\omega_2}, X_2) = \{6, 7\}$, $UPP(AS_{\omega_2}, X_2) =$

---

[2] The uncertainty and rough inclusion functions are defined as in Chap. 5. The distance measure is defined as follows $d(x, y) = |a(x) - a(y)|$.

$\{1, 2, 6, 7\}$, $LOW(AS_{\omega_{(1,2)}}, X_1 \times X_2) = \{1, 2, 5, 6\} \times \{6, 7\}$, $UPP(AS_{\omega_{(1,2)}}, X_1) = \{1, \dots, 6\} \times \{1, 2, 6, 7\}$. We also have $\pi_1(LOW(AS_{\omega_{(1,2)}}, X_1 \times X_2)) = \{1, 2, 5, 6\}$, $\pi_1(UPP(AS_{\omega_{(1,2)}}, X_1 \times X_2)) = \{1, \dots, 6\}$, $\pi_2(LOW(AS_{\omega_{(1,2)}}, X_1 \times X_2)) = \{6, 7\}$, $\pi_2(UPP(AS_{\omega_{(1,2)}}, X_1 \times X_2)) = \{1, 2, 6, 7\}$.

The compound approximation space corresponding to m database tables is defined as follows.

Definition 9.4 (*Compound approximation space $AS_{\omega_{(m)}}$*) Let $AS_{\omega_i} = (U_i, I_{\omega_i}, \nu_{\omega_i})$, where $1 \le i \le m$ and $m > 1$, be approximation spaces for information systems $IS_i = (U_i, A_i)$. A compound approximation space $AS_{\omega_{(m)}}$ for a compound information system $IS_{(m)} = \times(IS_i, \dots, IS_m)$ is defined by

$$AS_{\omega_{(m)}} = \times(AS_{\omega_1}, \dots, AS_{\omega_m}) = (U_{\omega_{(m)}}, I_{\omega_{(m)}}, \nu_{\omega_{(m)}}) \qquad (9.3)$$

where

- $U_{\omega_{(m)}} = \prod_{i=1}^{m} U_i$,

- $\underset{(x_1, \dots, x_m) \in U_{\omega_{(m)}}}{\forall} I_{\omega_{(m)}}((x_1, \dots, x_m)) = \prod_{i=1}^{m} I_{\omega_i}(x_i)$,

- $\underset{X_i, Y_i \in U_i, 1 \le i \le m}{\forall} \nu_{\omega_{(m)}}(\prod_{i=1}^{m} X_i, \prod_{i=1}^{m} Y_i) = \prod_{i=1}^{m} \nu_{\omega_i}(X_i, Y_i)$.

Proposition 9.3 *Let $AS_{\omega_{(m)}}$ be a CAS such that $\nu_1 = \dots = \nu_m = \nu_{SRI}$. The following holds*

$$\underset{X_i, Y_i \in U_i, 1 \le i \le m}{\forall} \nu_{\omega_{(i,\dots,m)}}(\prod_{i=1}^{m} X_i, \prod_{i=1}^{m} Y_i) = \nu_{SRI}(\prod_{i=1}^{m} X_i, \prod_{i=1}^{m} Y_i).$$

This can be proven analogously to Proposition 9.1.

Any subspace of $AS_{\omega_{(m)}}$ can be treated as a compound approximation space. More formally.

Proposition 9.4 *If $AS_{\omega_{(m)}}$ is a CAS, then so is $AS_{\omega_{(i_1, \dots, i_k)}}$, where $\{i_1, \dots, i_k\} \subseteq \{1, \dots, m\}$.*

This can be proven straightforwardly from Definition 9.4.

Approximations of a set in a compound approximation space are defined as follows.

Definition 9.5 (*Approximations of a set in $AS_{\omega_{(m)}}$*) Let $AS_{\omega_{(m)}} = (U_{\omega_{(m)}}, I_{\omega_{(m)}}, \nu_{\omega_{(m)}})$ be a CAS and $X_i \subseteq U_i$, where $1 \le i \le m$. The lower and upper approximations of the set $\prod_{i=1}^{m} X_i$ in $AS_{\omega_{(m)}}$ are defined, respectively, by

$$LOW(AS_{\omega_{(m)}}, \prod_{i=1}^{m} X_i) = \{(x_1, \dots, x_m) \in U_{\omega_{(m)}} : \nu_{\omega_{(m)}}(I_{\omega_{(m)}}((x_1, \dots, x_m)), \prod_{i=1}^{m} X_i) = 1\},$$

$$UPP(AS_{\omega_{(m)}}, \prod_{i=1}^{m} X_i) = \{(x_1, \ldots, x_m) \in U_{\omega_{(m)}} : v_{\omega_{(m)}}(I_{\omega_{(m)}}((x_1, \ldots, x_m)), \prod_{i=1}^{m} X_i) > 0\}.$$

The below proposition is a generalization of Proposition 9.2 and can be proven in an analogous way.

Proposition 9.5 *Let* $AS_{\omega_{(m)}}$ *be a CAS. The following hold* $\underset{X_i \subseteq U_i, 1 \leq i \leq m}{\forall}$

1. $$\underset{1 \leq i \leq j \leq m}{\forall} \prod_{l_1=1}^{i-1} U_{l_1} \times LOW(AS_{\omega_{(i,\ldots,j)}}, \prod_{l_2=i}^{j} X_{l_2}) \times \prod_{l_3=j+1}^{m} U_{l_3} = LOW(AS_{\omega_{(m)}}, \prod_{l_1=1}^{i-1} U_{l_1}$$
$$\times \prod_{l_2=i}^{j} X_{l_2} \times \prod_{l_3=j+1}^{m} U_{l_3}),$$

2. $$\underset{1 \leq i \leq j \leq m}{\forall} \prod_{l_1=1}^{i-1} U_{l_1} \times UPP(AS_{\omega_{(i,\ldots,j)}}, \prod_{l_2=i}^{j} X_{l_2}) \times \prod_{l_3=j+1}^{m} U_{l_3} = UPP(AS_{\omega_{(m)}}, \prod_{l_1=1}^{i-1} U_{l_1} \times$$
$$\prod_{l_2=i}^{j} X_{l_2} \times \prod_{l_3=j+1}^{m} U_{l_3}),$$

3. $$\underset{\{i_1,\ldots,i_k\} \subseteq \{1,\ldots,m\}}{\forall} LOW(AS_{\omega_{(i_1,\ldots,i_k)}}, \prod_{l=1}^{k} X_{i_l}) \neq \emptyset \Rightarrow \pi_{j_1,\ldots,j_{k'}}(LOW(AS_{\omega_{(m)}}, \prod_{i=1}^{m} X_i))$$
$$= LOW(AS_{\omega_{(j_1,\ldots,j_{k'})}}, \prod_{l=1}^{k'} X_{i_l}), \text{ where } \{j_1, \ldots, j_{k'}\} = \{1, \ldots, m\} \setminus \{i_1, \ldots, i_k\}.$$

4. $$\underset{\{i_1,\ldots,i_k\} \subseteq \{1,\ldots,m\}}{\forall} UPP(AS_{\omega_{(i_1,\ldots,i_k)}}, \prod_{l=1}^{k} X_{i_l}) \neq \emptyset \Rightarrow \pi_{j_1,\ldots,j_{k'}}(UPP(AS_{\omega_{(m)}}, \prod_{i=1}^{m} X_i))$$
$$= UPP(AS_{\omega_{(j_1,\ldots,j_{k'})}}, \prod_{l=1}^{k'} X_{i_l}), \text{ where } \{j_1, \ldots, j_{k'}\} = \{1, \ldots, m\} \setminus \{i_1, \ldots, i_k\}.$$

5. $$\prod_{i=1}^{m} LOW(AS_{\omega_i}, X_i) = LOW(AS_{\omega_{(m)}}, \prod_{i=1}^{m} X_i),$$

6. $$\prod_{i=1}^{m} UPP(AS_{\omega_i}, X_i) = UPP(AS_{\omega_{(m)}}, \prod_{i=1}^{m} X_i).$$

The following example illustrates the above definitions and propositions.

Example 9.2 Consider approximation spaces $AS_{\omega_{(3)}} = \times(AS_{\omega_1}, AS_{\omega_2}, AS_{\omega_3})$, where AS_{ω_1} and AS_{ω_2} are defined as in Example 9.1, and AS_{ω_3} is constructed based on relation *product* from Example 2.1. Let $I_{\omega_3} = I_{\{price\},\varepsilon_3}$, $\varepsilon_3 = (\varepsilon_{price}) = (1.00)$ and $v_{\omega_3} = v_{SRI}$. Let $X_1 = \{1, 2, 4, 7\} \subset U_1$, $X_2 = \{2, 6, 7\} \subset U_2$, $X_3 = \{2, 3, 4\} \subset U_3$, and $X_1 \times X_2 \times X_3 \subset U_{\omega_{(3)}}$ be the sets to be approximated.

The table from Example 9.1 is extended by the following columns. We obtain the following approximations $LOW(AS_{\omega_1}, X_1) = \{1, 2, 5, 6\}$, $UPP(AS_{\omega_1}, X_1) = \{1, \ldots, 6\}$, $LOW(AS_{\omega_2}, X_2) = \{6, 7\}$, $UPP(AS_{\omega_2}, X_2) = \{1, 2, 6, 7\}$, $LOW(AS_{\omega_3}, X_3) = \{2\}$, $UPP(AS_{\omega_3}, X_3) = \{1, 2, 3, 4, 5\}$.

The following illustrates the propositions formulated above.

1. We have $v_{\omega_{(3)}}(I_{\omega_1}(x_1) \times I_{\omega_2}(x_2) \times I_{\omega_3}(x_3), X_1 \times X_2 \times X_3) = v_{SRI}(I_{\omega_1}(x_1), X_1) \cdot v_{u,l}(I_{\omega_2}(x_2), X_2) \cdot v_{SRI}(I_{\omega_3}(x_3), X_3) = v_{SRI}(I_{\omega_1}(x_1) \times I_{\omega_3}(x_3), X_1 \times X_3) \cdot$

$x_3 \in U_3$	$I_{\omega_3}(x_3)$	$v_{\omega_3}(I_{\omega_3}(x_3), X_3)$
1	$\{1, 3\}$	0.5
2	$\{2, 3\}$	1
3	$\{1, 2, 3\}$	0.66
4	$\{4, 5\}$	0.5
5	$\{4, 5, 6\}$	0.33
6	$\{5, 6\}$	0

$v_{u,l}(I_{\omega_2}(x_2), X_2)$. For example, $v_{SRI}(I_{\omega_1}(3), X_1) \cdot v_{SRI}(I_{\omega_3}(5), X_3) = 0.5 \cdot 0.33 =$
0.17   and $v_{SRI}(I_{\omega_1}(3)$ $\times$ $I_{\omega_3}(5), X_1$ $\times$ $X_3)$ $=$
$\frac{card((\{3,4\} \times \{4,5,6\}) \cap (\{1,2,4,7\} \times \{2,3,4\}))}{card(\{3,4\} \times \{4,5,6\})} = 0.17$ (see Propositions 9.1 and 9.3).

2. The following subspaces of $AS_{\omega_{(3)}}$ are compound approximation spaces: $AS_{\omega_{(i,j)}}$, ($i \neq j$ and $i, j = 1, 2, 3$), AS_{ω_i} ($i = 1, 2, 3$). The last three ones can be called trivial compound approximation spaces (see Proposition 9.4).

3. The lower approximation of $X_1 \times X_2 \times X_3$ is $LOW(AS_{\omega_{(3)}}, X_1 \times X_2 \times X_3) =$
$\{(x_1, x_2, x_3) \in U_1 \times U_2 \times U_3 : v_{\omega_{(3)}}(I_{\omega_{(3)}}((x_1, x_2, x_3)), X_1 \times X_2 \times X_3) =$
$1\} = \{(x_1, x_2, x_3) \in U_1 \times U_2 \times U_3 : v_{\omega_1}(I_{\omega_1}(x_1), X_1) \cdot v_{\omega_2}(I_{\omega_2}(x_2), X_2) \cdot$
$v_{\omega_3}(I_{\omega_3}(x_3), X_3) = 1\} = \{1, 2, 5, 6\} \times \{6, 7\} \times \{2\} = LOW(AS_{\omega_1}, X_1) \times$
$LOW(AS_{\omega_2}, X_2) \times LOW(AS_{\omega_3}, X_3)$ (see Definition 9.5 and Proposition 9.5(5)).

4. We have $U_1 \times LOW(AS_{\omega_2}, X_2) \times U_3 = U_1 \times \{6, 7\} \times U_3 = LOW(AS_{\omega_{(3)}}, U_1 \times X_2 \times U_3)$ (see Proposition 9.5(1)).

5. We have $\pi_{1,2}(LOW(AS_{\omega_{(3)}}, X_1 \times X_2 \times X_3)) = \{1, 2, 5, 6\} \times \{6, 7\} = LOW(AS_{\omega_{(1,2)}}, X_1 \times X_2)$ (see Proposition 9.5(3)).

9.2.2   Constrained Compound Approximation Spaces

Analogously to compound approximation spaces, their constrained versions are defined for constrained compound information systems.

Firstly, the generalized theta-join operation on any subsets of the particular universes of a constrained compound information system is defined.

Definition 9.6 (*Operation $\bowtie_\Theta$ on subsets of universes*) Let $IS^\Theta_{(i,j)} = (U_i \bowtie_\Theta U_j, A_i \cup A_j)$ be a constrained compound information system. The operation $\bowtie_\Theta$ on subsets $A \subseteq U_i$ and $B \subseteq U_j$ is defined by

$$A \bowtie_\Theta B = \{(x_1, x_2) \in U_i \bowtie_\Theta U_j : (x_1, x_2) \in A \times B\}. \tag{9.4}$$

The $\bowtie_\Theta$ operation inherits properties of the Cartesian product.

Proposition 9.6 *Let $IS^\Theta_{(i,j)} = (U_i \bowtie_\Theta U_j, A_i \cup A_j)$ be a constrained compound information system. The following hold* $\underset{A,B \subseteq U_i; C,D \subseteq U_j}{\forall}$

1. $A \bowtie_\Theta \emptyset = \emptyset \bowtie_\Theta A = \emptyset$,
2. $A \bowtie_\Theta B \subseteq A \times B$,
3. $A \bowtie_\Theta (B \diamond C) = (A \bowtie_\Theta B) \diamond (C \bowtie_\Theta D)$, where $\diamond \in \{\cup, \cap, \backslash\}$,
4. $(A \diamond B) \bowtie_\Theta C = (A \bowtie_\Theta C) \diamond (B \bowtie_\Theta C)$, where $\diamond \in \{\cup, \cap, \backslash\}$,
5. $(A \cap B) \bowtie_\Theta (C \cap D) = (A \bowtie_\Theta C) \cap (B \bowtie_\Theta D)$,
6. $A \subseteq B \Rightarrow A \bowtie_\Theta C \subseteq B \bowtie_\Theta C$,
7. $C \subseteq D \Rightarrow A \bowtie_\Theta C \subseteq A \bowtie_\Theta D$,
8. $A \subseteq C \wedge B \subseteq D \Rightarrow A \bowtie_\Theta B \subseteq C \bowtie_\Theta D$.

Properties 2, 6–8 are restricted to the implications compared with the corresponding Cartesian product properties. Property 2 corresponds to $A \times B \equiv A \times B$.

The constrained compound approximation space corresponding two database tables is defined as follows.

Definition 9.7 (*Constrained compound approximation space* $AS^\Theta_{\omega_{(i,j)}}$) Let $AS_{\omega_i} = (U_i, I_{\omega_i}, v_{\omega_i})$ and $AS_{\omega_j} = (U_j, I_{\omega_j}, v_{\omega_j})$ be approximation spaces for information systems $IS_i = (U_i, A_i)$ and $IS_j = (U_j, A_j)$, respectively, where $v_\$ = v_{\omega_i} = v_{\omega_j}$. Let also $\Theta = \{\theta_1, \theta_2, \ldots \theta_n\} \in L_{IS_{i \wedge j}}$ be a set of joins of AS_{ω_i} and AS_{ω_j}. A constrained compound approximation space (CCAS for short) $AS^\Theta_{(i,j)}$ for a compound information system $IS^\Theta_{(i,j)} =\bowtie_\Theta (IS_i, IS_j)$ is defined by

$$AS^\Theta_{\omega_{(i,j)}} =\Rightarrow_\Theta (AS_{\omega_i}, AS_{\omega_j}) = (U^\Theta_{\omega_{(i,j)}}, I^\Theta_{\omega_{(i,j)}}, v^\Theta_{\omega_{(i,j)}}) \qquad (9.5)$$

where

- $U^\Theta_{\omega_{(i,j)}} = U_i \bowtie_\Theta U_j$,
- $\displaystyle \forall_{(x_1,x_2) \in U_i \bowtie_\Theta U_j} I^\Theta_{\omega_{(i,j)}}((x_1, x_2)) = I_{\omega_i}(x_1) \bowtie_\Theta I_{\omega_j}(x_2)$,
- $\displaystyle \forall_{X_1,Y_1 \in U_i, X_2,Y_2 \in U_j} v^\Theta_{\omega_{(i,j)}}(X_1 \bowtie_\Theta X_2, Y_1 \bowtie_\Theta Y_2) = v_\$(X_1 \bowtie_\Theta X_2, Y_1 \bowtie_\Theta Y_2)$.

The above definition is analogous to that of compound approximation space except that the rough inclusion function is simplified. Namely, one cannot use a different function for each universe, since the computation of inclusion degrees cannot be done separately. The reason is that the information about the connection of the universes is needed during computing the inclusion degrees.

The below proposition shows relationships between a CAS and its constrained version.

Proposition 9.7 *Let $AS_{\omega_{(i,j)}}$ and $AS^\Theta_{\omega_{(i,j)}}$ be a CAS and CCAS, respectively. The following holds*

1. $\displaystyle \forall_{(x_1,x_2) \in U^\Theta_{\omega_{(i,j)}}} I^\Theta_{\omega_{(i,j)}}((x_1, x_2)) = I_{\omega_{(i,j)}}((x_1, x_2)) \cap U^\Theta_{\omega_{(i,j)}}$.

2. $\displaystyle \forall_{X_1,Y_1 \in U_i, X_2,Y_2 \in U_j} X_1 \bowtie_\Theta X_2 = X_1 \times X_2 \wedge Y_1 \bowtie_\Theta Y_2 = Y_1 \times Y_2 \Rightarrow v^\Theta_{\omega_{(i,j)}}(X_1 \bowtie_\Theta X_2, Y_1 \bowtie_\Theta Y_2) = v_{\omega_{(i,j)}}(X_1 \times X_2, Y_1 \times Y_2)$.

Approximations of a set in a constrained compound approximation space are defined as follows.

Definition 9.8 (*Approximations of a set in* $AS^{\Theta}_{\omega_{(i,j)}}$) Let $AS^{\Theta}_{\omega_{(i,j)}} = (U^{\Theta}_{\omega_{(i,j)}}, I^{\Theta}_{\omega_{(i,j)}},$ $v^{\Theta}_{\omega_{(i,j)}})$ be a CCAS and $X_1 \subseteq U_i$, $X_2 \subseteq U_j$. The lower and upper approximations of the set $X_1 \bowtie_{\Theta} X_2 \subseteq U^{\Theta}_{\omega_{(i,j)}}$ in $AS^{\Theta}_{\omega_{(i,j)}}$ are defined, respectively, by

$$LOW(AS^{\Theta}_{\omega_{(i,j)}}, X_1 \bowtie_{\Theta} X_2) = \{(x_1, x_2) \in U^{\Theta}_{\omega_{(i,j)}} : v^{\Theta}_{\omega_{(i,j)}}(I^{\Theta}_{\omega_{(i,j)}}((x_1, x_2)), X_1 \bowtie_{\Theta} X_2) = 1\},$$

$$UPP(AS^{\Theta}_{\omega_{(i,j)}}, X_1 \bowtie_{\Theta} X_2) = \{(x_1, x_2) \in U^{\Theta}_{\omega_{(i,j)}} : v^{\Theta}_{\omega_{(i,j)}}(I^{\Theta}_{\omega_{(i,j)}}((x_1, x_2)), X_1 \bowtie_{\Theta} X_2) > 0\}.$$

The following two propositions show properties of the lower and upper approximations and they correspond to those from Proposition 9.5.

Proposition 9.8 *Let* $AS_{\omega_{(i,j)}}$ *be a CAS and* $AS^{\Theta}_{\omega_{(i,j)}}$ *its constrained version such that* $v_{\omega_i}, v_{\omega_j}$ *and* $v^{\Theta}_{\omega_{(i,j)}}$ *are RIFs. The following hold* $\underset{X_1 \subseteq U_1, X_2 \subseteq U_2}{\forall}$

1. $\underset{(x_1,x_2) \in U^{\Theta}_{\omega_{(i,j)}}}{\forall} I^{\Theta}_{\omega_{(i,j)}}((x_1, x_2)) = I_{\omega_{(i,j)}}((x_1, x_2)) \Rightarrow$

    a. $LOW(AS^{\Theta}_{\omega_{(i,j)}}, X_1 \bowtie_{\Theta} X_2) \subseteq LOW(AS_{\omega_{(i,j)}}, X_1 \times X_2)),$

    b. $\pi_k(LOW(AS^{\Theta}_{\omega_{(i,j)}}, X_1 \bowtie_{\Theta} X_2)) \subseteq LOW(AS_{\omega_k}, X_k),$ *where* $k = 1, 2$.

2. $LOW(AS_{\omega_i}, X_1) \bowtie_{\Theta} U_j \subseteq LOW(AS^{\Theta}_{\omega_{(i,j)}}, X_1 \bowtie_{\Theta} U_j),$

3. $LOW(AS_{\omega_i}, X_1) \bowtie_{\Theta} LOW(AS_{\omega_j}, X_2) \subseteq LOW(AS^{\Theta}_{\omega_{(i,j)}}, X_1 \bowtie_{\Theta} X_2).$

For AS_{ω_j} one can formulate an equality analogous to 2.

Proposition 9.9 *Let* $AS_{\omega_{(i,j)}}$ *be a CAS and* $AS^{\Theta}_{\omega_{(i,j)}}$ *its constrained version such that* $v_{\omega_i}, v_{\omega_j}$ *and* $v^{\Theta}_{\omega_{(i,j)}}$ *satisfy property* p_5. *The following hold* $\underset{X_1 \subseteq U_1, X_2 \subseteq U_2}{\forall}$

1. $UPP(AS^{\Theta}_{\omega_{(i,j)}}, X_1 \bowtie_{\Theta} X_2) \subseteq UPP(AS_{\omega_{(i,j)}}, X_1 \times X_2)),$

2. $\pi_k(UPP(AS^{\Theta}_{\omega_{(i,j)}}, X_1 \bowtie_{\Theta} X_2)) \subseteq UPP(AS_{\omega_k}, X_k),$ *where* $k = 1, 2$,

3. $UPP(AS_{\omega_i}, X_1) \bowtie_{\Theta} U_j = UPP(AS^{\Theta}_{\omega_{(i,j)}}, X_1 \bowtie_{\Theta} U_j).$

4. $UPP(AS_{\omega_i}, X_1) \bowtie_{\Theta} UPP(AS_{\omega_j}, X_2) = UPP(AS^{\Theta}_{\omega_{(i,j)}}, X_1 \bowtie_{\Theta} X_2).$

For AS_{ω_j} one can formulate an equality analogous to 3.

It is worth stressing that the equalities from Proposition 9.2 do not hold in general for a constrained compound approximation space. It means that rough sets derived from the approximations from Propositions 9.8 and 9.9 have a different meaning than those of a compound approximation space. Therefore, applying a constrained version of the compound approximation space one can derive new knowledge. This issue will be studied in the next section.

Example 9.3 Consider a CCAS $AS^{\Theta}_{\omega_{(1,2)}} = \bowtie_{\Theta} (AS_{\omega_1}, AS_{\omega_2})$ where $AS_{\omega_1}, AS_{\omega_2}$ are defined as in Example 9.1, $\Theta = \{\theta\}$, $\theta = (customer.id, purchase.cust\_id)$, and $U^{\Theta}_{\omega_{(1,2)}} = \{(1,1), (1,2), (2,3), (2,4), (3,8), (4,5), (4,6), (6,7)\}$.
Let $X_1 = \{1, 2, 4, 7\} \subset U_1$ and $X_2 = \{2, 6, 7\} \subset U_2$. We will approximate the set $X_1 \bowtie_{\Theta} X_2 = \{(1,2), (4,6)\}$.
The table below shows the similarity classes and their rough inclusion degrees in the respective sets.

$(x_1, x_2) \in U_1 \bowtie_{\Theta} U_2$	$I^{\Theta}_{\omega_{(1,2)}}((x_1, x_2))$	$v_{\omega_{(1,2)}}(I^{\Theta}_{\omega_{(1,2)}}((x_1, x_2)), X_1 \bowtie_{\Theta} X_2)$
(1, 1)	$\{(1,1), (1,2)\}$	0.5
(1, 2)	$\{(1,1), (1,2)\}$	0.5
(2, 3)	$\{(2,3), (2,4)\}$	0
(2, 4)	$\{(2,3), (2,4)\}$	0
(3, 8)	$\{(4,5), (3,8)\}$	0
(4, 5)	$\{(4,5), (3,8)\}$	0
(4, 6)	$\{(4,6)\}$	1
(6, 7)	$\{(6,7)\}$	0

We obtain the following approximations $LOW(AS^{\Theta}_{\omega_{(1,2)}}, X_1 \bowtie_{\Theta} X_2) = \{(4,6)\}$ and $UPP(AS^{\Theta}_{\omega_{(1,2)}}, X_1 \bowtie_{\Theta} X_2) = \{(1,1), (1,2), (2,4)\}$.
We also have $\pi_1(LOW(AS^{\Theta}_{\omega_{(1,2)}}, X_1 \bowtie_{\Theta} X_2)) = \{4\}, \pi_1(UPP(AS^{\Theta}_{\omega_{(1,2)}}, X_1 \bowtie_{\Theta} X_2)) = \{1, 2\}, \pi_2(LOW(AS^{\Theta}_{\omega_{(1,2)}}, X_1 \bowtie_{\Theta} X_2)) = \{6\}$ and $\pi_2(UPP(AS^{\Theta}_{\omega_{(1,2)}}, X_1 \bowtie_{\Theta} X_2)) = \{1, 2, 4\}$.
For this approximation space the assumption of Proposition 9.8 (1) is not satisfied. Therefore, we have $\pi_1(LOW(AS^{\Theta}_{\omega_{(1,2)}}, X_1 \bowtie_{\Theta} X_2)) = \{4\} \nsubseteq LOW(AS_{\omega_1}, X_1) = \{1, 2, 5, 6\}$.

The constrained compound approximation space corresponding to m database tables is defined as follows.

Definition 9.9 (*Constrained compound approximation space $AS^{\Theta}_{\omega_{(m)}}$*) Let $AS_{\omega_i} = (U_i, I_{\omega_i}, v_{\omega_i})$ be approximation spaces for information systems $IS_i = (U_i, A_i)$, and $v_{\$} = v_{\omega_i}$ for $1 \leq i \leq m$. Let also $\Theta = \{\theta_1, \theta_2, \ldots \theta_n\}$ be a set of joins of AS_{ω_i} such that

$$\underset{1 < j \leq m}{\forall} \underset{i < j}{\exists} U_i \bowtie_{\Theta} U_j \neq \emptyset \text{ (each approximation space joins with some earlier con-}$$
sidered space).
A constrained compound approximation space $AS^{\Theta}_{\omega_{(m)}}$ for a constrained compound information system $IS^{\Theta}_{(m)}$ is defined by

$$AS^{\Theta}_{\omega_{(m)}} = \bowtie_{\Theta} (AS_{\omega_1}, \ldots, AS_{\omega_m}) = (U^{\Theta}_{\omega_{(m)}}, I^{\Theta}_{\omega_{(m)}}, v^{\Theta}_{\omega_{(m)}}) \qquad (9.6)$$

where

- $U^{\Theta}_{\omega_{(m)}} = U_1 \bowtie_{\Theta} \ldots \bowtie_{\Theta} U_m,$

- $$\underset{(x_1,\ldots,x_m)\in U_1\bowtie_\Theta\ldots\bowtie_\Theta U_m}{\forall} I^\Theta_{\omega_{(m)}}((x_1,\ldots,x_m)) = I_{\omega_1}(x_1)\bowtie_\Theta\ldots\bowtie_\Theta I_{\omega_m}(x_m),$$
- $$\underset{X_i,Y_i\in U_i, 1\le i\le m}{\forall} v^\Theta_{\omega_{(m)}}(X_1\bowtie_\Theta\ldots\bowtie_\Theta X_m, Y_1\bowtie_\Theta\ldots\bowtie_\Theta Y_m) = v_\$(X_1\bowtie_\Theta$$
$$\ldots\bowtie_\Theta X_m, Y_1\bowtie_\Theta\ldots\bowtie_\Theta Y_m).$$

Analogously to $AS_{\omega_{(m)}}$ we can consider any subspace of $AS^\Theta_{\omega_{(m)}}$.

Proposition 9.10 *If $AS^\Theta_{\omega_{(m)}}$ is a CCAS, then so is $AS^{\Theta'}_{\omega_{(i_1,\ldots,i_k)}}$, where $\{i_1,\ldots,i_k\}\subseteq$*
$\{1,\ldots,m\}$, $\Theta'\subseteq\Theta$ and $\underset{i_1<j\le i_k}{\forall}\underset{i<j}{\exists} U_i\bowtie_{\Theta'} U_j\ne\emptyset$.

This can be proven straightforwardly from Definition 9.9.

Approximations of a set in a constrained compound approximation space are defined as follows.

Definition 9.10 (*Approximations of a set in $AS^\Theta_{\omega_{(m)}}$*) Let $AS^\Theta_{\omega_{(m)}} = (U^\Theta_{\omega_{(m)}}, I^\Theta_{\omega_{(m)}}, v^\Theta_{\omega_{(m)}})$ be a CCAS and $X_i\subseteq U_i$, where $1\le i\le m$. The lower and upper approximations of the set $X_1\bowtie_\Theta\ldots\bowtie_\Theta X_m$ in $AS^\Theta_{\omega_{(m)}}$ are defined, respectively, by

$LOW(AS^\Theta_{\omega_{(m)}}, X_1\bowtie_\Theta\ldots\bowtie_\Theta X_m) = \{(x_1,\ldots,x_m)\in U^\Theta_{\omega_{(m)}} : v^\Theta_{\omega_{(m)}}(I^\Theta_{\omega_{(m)}}((x_1,\ldots,x_m)), X_1\bowtie_\Theta$
$\ldots\bowtie_\Theta X_m) = 1\}$,

$UPP(AS^\Theta_{\omega_{(m)}}, X_1\bowtie_\Theta\ldots\bowtie_\Theta X_m) = \{(x_1,\ldots,x_m)\in U^\Theta_{\omega_{(m)}} : v^\Theta_{\omega_{(m)}}(I^\Theta_{\omega_{(m)}}((x_1,\ldots,x_m)), X_1\bowtie_\Theta$
$\ldots\bowtie_\Theta X_m) > 0\}$.

For any constrained compound approximation space $AS^\Theta_{\omega_{(m)}}$ one can formulate propositions analogous to Propositions 9.7, 9.8, and 9.9.

9.3 Knowledge Derived from Approximations of Compound Concepts

This sections examines the construction of concepts and their restricted versions in compound approximation spaces, i.e. compound concepts. This also shows what knowledge can be derived from the database using approximations of the concepts.

9.3.1 Compound Concepts

Theoretically the concept to be approximated can be defined by any subset of the universe. In practice, we consider a subset of objects that share the same feature (usually a decision class). We will limit our discussion to concepts that can be identified by features occurring in the database.

Definition 9.11 (*Compound concept*) Let $AS_{\omega_{(m)}}$ be a CAS and $\underset{1\le j\le k}{\forall}\alpha_{i_j}\in L_{IS_{i_j}}\vee$
$\alpha_{i_j} = \emptyset$, where $\{i_1,\ldots,i_k\}\in\{1,\ldots,m\}$.

A subset $\prod_{j=1}^{k} U^{\alpha_{i_k}} = SEM_{IS_{(i_1,\dots,i_k)}}(\bigwedge_{j=1}^{k} \alpha_{i_j}) \subseteq U_{\omega_{(i_1,\dots,i_k)}}$ is a compound concept in $AS_{\omega_{(m)}}$.

A compound concept in $AS^{\Theta}_{\omega_{(m)}}$ can be defined analogously.

As mentioned in the previous section, compared with compound approximation spaces their constrained version make it possible to constructs approximations with a different meaning. Thanks to this, one can derive new knowledge from data. This will be illustrated using the following example.

Example 9.4 Consider the CCAS $AS^{\Theta}_{\omega_{(1,2)}} =\bowtie_{\Theta} (AS_{\omega_1}, AS_{\omega_2})$ from Example 9.3. We construct all possible sets using the lower approximation. This can be done analogously for the upper approximation and combinations of both (see Proposition 9.8).

1. Good customers.
   The concept is defined by $U_1^{\alpha_1} = \{1, 2, 4, 5, 6\}$ where $\alpha_1 = (class, 1)$.
   We obtain $LOW(AS_{\omega_1}, U^{\alpha_1}) = \{1, 2, 5, 6\}$ (certainly good customers).
2. Small purchases (one piece).[3]
   The concept is defined by $U_2^{\alpha_2} = \{1, 3, 4, 5, 8\}$ where $\alpha_2 = (amount, 1)$.
   We obtain $LOW(AS_{\omega_2}, U^{\alpha_2}) = \{4, 5, 8\}$ (certainly small purchases).
3. Good customers and their purchases.
   The concept is defined by $U_1^{\alpha_1} \bowtie_{\Theta} U_2 = \{(1, 1), (1, 2), (2, 3), (2, 4), (4, 5),$
   $(4, 6), (6, 7)\}$.

   a. We obtain $S_1 = LOW(AS_{\omega_1}, U_1^{\alpha_1}) \bowtie_{\Theta} U_2 = \{(1, 1), (1, 2), (2, 3), (2, 4),$
      $(6, 7)\}$, where $\pi_1(S_1) = \{1, 2, 6\}$ (certainly good customers who make purchases) and $\pi_2(S_1) = \{1, 2, 3, 4, 7\}$ (purchases made by certainly good customers).
   b. We obtain $S_2 = LOW(AS^{\Theta}_{\omega_{(1,2)}}, U_1^{\alpha_1} \bowtie_{\Theta} U_2) = \{(1, 1), (1, 2), (2, 3),$
      $(2, 4), (4, 6), (6, 7)\}$, where $\pi_1(S_2) = \{1, 2, 4, 6\}$ (good customers who make purchases that are certainly made by good customers) and $\pi_2(S_1) = \{1, 2, 3, 4, 6, 7\}$ (purchases that are certainly made by good customers).
      Customer 4 is included in $\pi_1(S_2)$ due to the pair $(4, 6)$ but she cannot be added to the set due to the pair $(4, 5)$. Namely, purchase 5 is not certainly made by a good customer, since $(4, 5)$ is similar to $(3, 8)$ and 3 is not a good customer.

4. Customers and their small purchases.
   The concept is defined by $U_1 \bowtie_{\Theta} U_2^{\alpha_2} = \{(1, 1), (2, 3), (2, 4), (3, 8), (4, 5)\}$.

   a. We obtain $S_3 = U_1 \bowtie_{\Theta} LOW(AS_{\omega_2}, U_2^{\alpha_2}) = \{(2, 4), (3, 8), (4, 5)\}$, where $\pi_1(S_3) = \{2, 3, 4\}$ (customers who make certainly small purchases) and $\pi_2(S_3) = \{4, 5, 8\}$ (certainly small purchases made by customers).
   b. We obtain $S_4 = LOW(AS^{\Theta}_{\omega_{(1,2)}}, U_1 \bowtie_{\Theta} U_2^{\alpha_2}) = \{(2, 3), (2, 4), (3, 8),$
      $(4, 5)\}$, where $\pi_1(S_4) = \{2, 3, 4\}$ (customers who certainly make small

---

[3]Here, a purchase is identified by one row in the *purchase* table.

purchases) and $\pi_2(S_4) = \{3, 4, 5, 8\}$ (purchases made by customers who certainly make small purchases).

5. Good customers and their small purchases.
   The concept is defined by $U_1^{\alpha_1} \bowtie_\Theta U_2^{\alpha_2} = \{(1, 1), (2, 3), (2, 4), (4, 5)\}$.

   a. We obtain $S_5 = LOW(AS_{\omega_1}, U_1^{\alpha_1}) \bowtie_\Theta U_2^{\alpha_2} = \{(1, 1), (2, 3), (2, 4)\}$, where $\pi_1(S_5) = \{1, 2\}$ (certainly good customers who make small purchases) and $\pi_2(S_5) = \{1, 3, 4\}$ (small purchases made by certainly good customers).

   b. We obtain $S_6 = U_1^{\alpha_1} \bowtie_\Theta LOW(AS_{\omega_2}, U_2^{\alpha_2}) = \{(4, 5)\}$, where $\pi_1(S_6) = \{4\}$ (good customers who make certainly small purchases) and $\pi_2(S_6) = \{5\}$ (certainly small purchases made by good customers).

   c. We obtain $S_7 = LOW(AS_{\omega_1}, U_1^{\alpha_1}) \bowtie_\Theta LOW(AS_{\omega_2}, U_2^{\alpha_2}) = \{(2, 4)\}$, where $\pi_1(S_7) = \{2\}$ (certainly good customers who make certainly small purchases) and $\pi_2(S_7) = \{4\}$ (certainly small purchases made by certainly good customers).

   d. We obtain $S_8 = LOW(AS_{\omega_{(1,2)}}^\Theta, U_1^{\alpha_1} \bowtie_\Theta U_2^{\alpha_2}) = \{(2, 3), (2, 4)\}$, where $\pi_1(S_8) = \{2\}$ (good customers who certainly make small purchases) and $\pi_2(S_8) = \{3, 4\}$ (small purchases certainly made by good customers).
   Customer 1 is not included in $\pi_1(S_8)$ because one of his purchases, i.e. 2, is not small. In turn, purchase 5 is not included in $\pi_2(S_8)$ because the pair (4, 5) is similar to (3, 8) and customer 3 is not good.

In points 3a, 4a, and 5a–5c we obtain new knowledge compared with 3b, 4b, and 5d, respectively.

9.3.2 Restricted Compound Concepts

Based on a compound concept one can construct its restricted version by considering selected universes of a given CCAS.

Definition 9.12 (*Restricted concept*) Let $AS_{\omega_{(m)}}^\Theta$ be a CCAS. A concept restricted by $\{i_1, \ldots, i_k\} \subseteq \{1, \ldots, m\}$ with respect to a concept $X_1 \bowtie_\Theta \ldots \bowtie_\Theta X_m \in U_{\omega_{(m)}}^\Theta$ is defined by

$$\pi_{i_1, \ldots, i_k}(X_1 \bowtie_\Theta \ldots \bowtie_\Theta X_m). \tag{9.7}$$

The below proposition shows that a restricted concept of a CCAS may differ from its corresponding concept defined in the subspace of the CCAS.

Proposition 9.11 Let $AS_{\omega_{(m)}}^\Theta$ be a CCAS, $\emptyset \neq X_1 \bowtie_\Theta \ldots \bowtie_\Theta X_m \in U_{\omega_{(m)}}^\Theta$ and $\{i_1, \ldots, i_k\} \subseteq \{1, \ldots, m\}$. The following hold

$$\begin{cases} \emptyset \neq \pi_{i_1, \ldots, i_k}(X_1 \bowtie_\Theta \ldots \bowtie_\Theta X_m) \subseteq X_{i_1} \bowtie_\Theta \ldots \bowtie_\Theta X_{i_k}, & \text{if } AS_{\omega_{(i_1, \ldots, i_k)}}^\Theta \text{ is a CCAS}; \\ X_1 \bowtie_\Theta \ldots \bowtie_\Theta X_m = \emptyset, & \text{otherwise.} \end{cases}$$

Approximations of restricted concepts are defined in the following way.

Definition 9.13 (*Approximations of restricted concepts in* $AS^{\Theta}_{\omega(i,j)}$) Let $AS^{\Theta}_{\omega(i,j)}$ be a CCAS and $X_1 \subseteq U_i$, $X_2 \subseteq U_j$. The lower and upper approximations of the restricted concept $\pi_k(X_1 \bowtie_\Theta X_2)$ $(k = 1, 2)$ in $AS^{\Theta}_{\omega(i,j)}$ are defined, respectively, by

$$LOW(AS^{\Theta}_{\omega(i,j)}, \pi_k(X_1 \bowtie_\Theta X_2)) = \pi_k(LOW(AS^{\Theta}_{\omega(i,j)}, X_1 \bowtie_\Theta X_2)),$$

$$UPP(AS^{\Theta}_{\omega(i,j)}, \pi_k(X_1 \bowtie_\Theta X_2)) = \pi_k(UPP(AS^{\Theta}_{\omega(i,j)}, X_1 \bowtie_\Theta X_2)).$$

The approximations of a restricted concept may change if we replace a CCAS with its subspace corresponding to the restricted concept. This difference is shown in the below proposition.

Proposition 9.12 *Let* $AS^{\Theta}_{\omega(i,j)} =\bowtie_\Theta (AS_{\omega_i}, AS_{\omega_j})$ *be a CCAS such that* v_{ω_i}, v_{ω_j} *and* $v_{\omega(i,j)}$ *are RIFs (and satisfy property* p_5[4] *(see Chap. 5)). The following hold* $\underset{X_1 \subseteq U_i}{\forall}$

1. $LOW(AS_{\omega_i}, \pi_i(X_1 \bowtie_\Theta U_j)) \subseteq \pi_i(LOW(AS^{\Theta}_{\omega(i,j)}, X_1 \bowtie_\Theta U_j))$,
2. $LOW(AS_{\omega_i}, \pi_i(X_1 \bowtie_\Theta U_j)) \subseteq \pi_i(LOW(AS_{\omega_i}, X_1) \bowtie_\Theta U_j)$,
3. $UPP(AS_{\omega_i}, \pi_i(X_1 \bowtie_\Theta U_j)) = \pi_i(UPP(AS^{\Theta}_{\omega(i,j)}, X_1 \bowtie_\Theta U_j))$,
4. $UPP(AS_{\omega_i}, \pi_i(X_1 \bowtie_\Theta U_j)) = \pi_i(UPP(AS_{\omega_i}, X_1) \bowtie_\Theta U_j)$.

For AS_{ω_j} one can formulate analogous equalities.

The following example illustrates the above proposition.

Example 9.5 Consider a CCAS $AS^{\Theta}_{\omega(1,2,3)} =\bowtie_\Theta (AS_{\omega_1}, AS_{\omega_2}, AS_{\omega_3})$ where $(AS_{\omega_1}$ and AS_{ω_2} are defined as in Example 9.3, and AS_{ω_3} is constructed based on the *product* table, $I_{\omega_3} = \emptyset$ (all objects are similar to one another). Consider a concept "Good customers, their small purchases and products bought" that is defined by $\mathcal{X} = U_1^{\alpha_1} \bowtie_\Theta U_2^{\alpha_2} \bowtie_\Theta U_3 = \{(1, 1, 1), (2, 3, 1), (2, 4, 3), (4, 5, 6)\}$ where $\alpha_1 = (class, 1)$ and $\alpha_2 = (amount, 1)$. We construct restricted concepts.

1. Good customers in $\mathcal{X}$.
   The concept is defined by $\pi_1(\mathcal{X}) = \{1, 2, 4\} \subset U_1^\alpha = \{1, 2, 4, 5, 6\}$.

   a. We obtain $LOW(AS_{\omega(3)}, \pi_1(\mathcal{X})) = \{2\}$ (good customers who certainly make small purchases).
   b. We have $LOW(AS_{\omega_1}, \pi_1(\mathcal{X})) = \{1, 2\}$ (certainly good customers of those who make small purchases). Customer number 1 is certainly good one but he does not certainly make small purchases due to his purchase number 2 that includes two pieces of product number 3.
   c. We obtain $LOW(AS_{\omega_1}, U_1^{\alpha_1}) \bowtie_\Theta U_2 = \{1, 2, 6\}$ (certainly good customers of those who make small purchases).

2. Good customers and their product in $\mathcal{X}$.
   The concept is defined by $\pi_{1,3}(\mathcal{X}) = \{(1, 1), (2, 1), (2, 3), (4, 6)\} \supset U_1^{\alpha_1} \bowtie_\Theta U_3^{u_2} = \{1, 2, 4, 5, 6\} \bowtie_\Theta U_3 = \emptyset$. We can only compute $LOW(AS_{\omega(3)}, \pi_{1,3}$

---

[4]The condition is required for the last two equalities.

$(\mathcal{X})) = \{(2, 1), (2, 3)\}$ (good customers and their products that are certainly purchased by them in a small amount).

Proposition 9.12 and Example 9.5 show that it is possible to obtain new knowledge from approximations when restricted concepts are considered.

9.4 Evaluation of the Approach

This section provides a complexity analysis of the approach.

We will start with evaluating the cost of computing approximations in non-compound approximation spaces. Let $n = card(U)$, where U is the universe in an approximation space $AS_\omega = (U, I_\omega, \nu_\omega)$.

- The cost of computing $I_\omega(x)$ for $x \in U$ is $T_1(n) = n = O(n)$.
- The cost of computing $\nu_\omega(X, Y)$ for $X, Y \subseteq U$ is $T_2(n) = card(X)card(Y) \le n^2 = O(n^2)$.
- The cost of computing $APP(AS_\omega, X)$ for $X \subseteq U$, where $APP \in \{LOW, UPP\}$ is $T_3(n) = n(T_1(n) + T_2(n)) = O(n^3)$.

Consider a compound approximation space $AS_{\omega(i,j)} = (U_{\omega(i,j)}, I_{\omega(i,j)}, \nu_{\omega(i,j)})$, where $U_{\omega(i,j)} = U_i \times U_j$. Let $n_i = card(U_i)$, $n_j = card(U_j)$, and $n^{max} = max\{n_i, n_j\}$.

- The cost of computing $I_{\omega(i,j)}((x_1, x_2))$ for $(x_1, x_2) \in U_{\omega(i,j)}$ is $T_1(n_i, n_j) \le T_1(n_i) + T_1(n_j) + n_i n_j = O((n^{max})^2)$. The sets $I_{\omega_i}(x_1)$ and $I_{\omega_j}(x_2)$ can be computed separately (see Definition 9.2). The cost of computing the Cartesian product of the both sets is up to $n_i n_j$.
- The cost of computing $\nu_{\omega(i,j)}(X1 \times X_2, Y_1 \times Y_2)$ for $X_1, Y_1 \subseteq U_i$ and $X_2, Y_2 \subseteq U_j$ is $T_2(n_i, n_j) = T_2(n_i) + T_2(n_j) + 1 = O((n^{max})^2)$, where "1" is the cost of multiplying $\nu_{\omega_i}(X_1, Y_1)$ by $\nu_{\omega_j}(X_2, Y_2)$.
- The cost of computing $APP(AS_{\omega(i,j)}, X_1 \times X_2)$ for $X_1 \subseteq U_i$ and $X_2 \subseteq U_j$ is $T_3(n_i, n_j) \le T_3(n_i) + T_3(n_j) + n_i n_j = O((n^{max})^3)$, where $n_i n_j$ is the maximal cost of computing the Cartesian product of $APP(AS_{\omega_i}, X_1)$ and $APP(AS_{\omega_j}, X_2)$ (see Proposition 9.2).

We analogously analyze the complexity for a compound approximation space $AS_{\omega(m)} = (U_{\omega(m)}, I_{\omega(m)}, \nu_{\omega(m)})$, where $U_{\omega(m)} = \prod_{i=1}^{m} U_i$. Let $n_i = card(U_i)$ and $n^{max} = max\{n_i : 1 \le i \le m\}$.

- The cost of computing $I_{\omega(m)}((x_1, \ldots, x_m))$ for $(x_1 \ldots, x_m) \in U_{\omega(m)}$ is $T_1(n_1, \ldots, n_m) \le \sum_{i=1}^{m} T_1(n_i) + \prod_{i=1}^{m} n_i = O((n^{max})^m)$. It is assumed that m is significantly lower than n and thereby it does not influence the cost.
- The cost of computing $\nu_{\omega(m)}(\prod_{i=1}^{m} X_i, \prod_{i=1}^{m} Y_i)$ for $X_i, Y_i \subseteq U_i$ is $T_2(n_1, \ldots, n_m) = \sum_{i=1}^{m} T_2(n_i) + m - 1 = O((n^{max})^2)$.

- The cost of computing $APP(AS_{\omega_{(m)}}, \prod_{i=1}^{m} X_i)$ for $X_i \subseteq U_i$ is $T_3(n_1, \ldots, n_m) \leq$

$$\sum_{i=1}^{m} T_3(n_i) + \prod_{i=1}^{m} n_i = O((n^{max})^3) + O((n^{max})^m) = O((n^{max})^k), \text{ where } k = max\{3, m\}.$$

It is worth noting that for $AS_{\omega_{(m)}}$ the cost of computing the value of the rough inclusion function does not increase compared with $AS_{\omega_{(i,j)}}$. For $AS_{\omega_{(m)}}$ this task can be divided into subtasks, each involving one particular universe. Since the partial results are numbers, the cost of joining them (i.e. multiplying them by one another) is slight.

Consider now a constrained compound approximation space $AS_{\omega_{(i,j)}}^{\Theta} = (U_{\omega_{(i,j)}}^{\Theta}, I_{\omega_{(i,j)}}^{\Theta}, v_{\omega_{(i,j)}}^{\Theta})$, where $U_{\omega_{(i,j)}}^{\Theta} = U_i \bowtie_{\Theta} U_j$.

- The cost of computing $I_{\omega_{(i,j)}}^{\Theta}((x_1, x_2))$ for $(x_1, x_2) \in U_{\omega_{(i,j)}}^{\Theta}$ is $T_{1'}(n_i, n_j) \leq T_1(n_i) + T_1(n_j) + n_i n_j = O((n^{max})^2)$.
- The cost of computing $v_{\omega_{(i,j)}}^{\Theta}(X1 \bowtie_{\Theta} X_2, Y_1 \bowtie_{\Theta} Y_2)$ for $X_1, Y_1 \subseteq U_i$ and $X_2, Y_2 \subseteq U_j$ is $T_{2'}(n_i, n_j) = card(X1 \bowtie_{\Theta} X_2)card(Y_1 \bowtie_{\Theta} Y_2) \leq ((n^{max})^2)^2 = O((n^{max})^4)$.
- The cost of computing $APP(AS_{\omega_{(i,j)}}^{\Theta}, X_1 \bowtie_{\Theta} X_2)$ for $X_1 \subseteq U_i$ and $X_2 \subseteq U_j$ is $T_{3'}(n_i, n_j) = card(U_{\omega_{(i,j)}}^{\Theta})(T_{1'}^{(2)}(n_i, n_j) + T_{2'}^{(2)}(n_i, n_j)) \leq O((n^{max})^6)$.

Finally, we analyze the complexity for a constrained compound approximation space $AS_{\omega_{(m)}}^{\Theta} = (U_{\omega_{(m)}}^{\Theta}, I_{\omega_{(m)}}^{\Theta}, v_{\omega_{(m)}}^{\Theta})$, where $U_{\omega_{(m)}}^{\Theta} = U_1 \bowtie_{\Theta} \ldots \bowtie_{\Theta} U_m$.

- The cost of computing $I_{\omega_{(m)}}^{\Theta}((x_1, \ldots, x_m))$ for $(x_1 \ldots, x_m) \in U_{\omega_{(m)}}^{\Theta}$ is

$$T_{1'}(n_1, \ldots, n_m) \leq \sum_{i=1}^{m} T_1(n_i) + \prod_{i=1}^{m} n_i = O((n^{max})^m).$$

- The cost of computing $v_{\omega_{(m)}}(\prod_{i=1}^{m} X_i, \prod_{i=1}^{m} Y_i)$ for $X_i, Y_i \subseteq U_i$ is $T_{2'}(n_1, \ldots, n_m) = card(X1 \bowtie_{\Theta} \ldots \bowtie_{\Theta} X_m)card(Y_1 \bowtie_{\Theta} \ldots \bowtie_{\Theta} Y_m) \leq ((n^{max})^m)^2 = O((n^{max})^{2m})$.
- The cost of computing $APP(AS_{\omega_{(m)}}, \prod_{i=1}^{m} X_i)$ for $X_i \subseteq U_i$ is $T_{3'}(n_1, \ldots, n_m) = card(U_{\omega_{(m)}}^{\Theta}) \left(T_{1'}^{(m)}(n_1, \ldots, n_m) + T_{2'}^{(m)}(n_1, \ldots, n_m) \right) = O((n^{max})^{3m})$.

The pessimistic complexity of operations for constrained compound approximation spaces is considerably higher. It applies to the case when the constrained compound universe is (almost) as big as its non-constrained equivalent. In practice, the former universe is significantly smaller thanks to the constraints defined by the theta-join and imposed on the latter universe.

9.5   Conclusions

This chapter has introduced compound approximation spaces and their constrained versions. They are defined for previously introduced compound information systems. Compound approximation spaces are viewed as extensions of tolerance approximation spaces to a relational case.

The main benefits of the approach are summarized as follows.

1. Tolerance rough set model used in the construction of a compound space enables to properly adapt rough set tools to each separate universe.
2. Compound approximations spaces make it possible to simultaneously approximate more than one separate concept, each concerning a different universe.
3. Constrained compound approximations spaces additionally enable to compute approximations of the relationship of two concepts.
4. A concept of a given universe can be more precisely specified and approximated by using its restricted version that makes it possible to use conditions concerning the whole compound approximation space.

Chapter 10
Conclusions

This monograph has outlined the state of the art of a newly emerging research area called granular-relational data mining. Two general approaches for constructing granular computing frameworks intended to mine relational data have comprehensively been described: generalized related set based approach and description language based approach.

The choice of the framework can be dependent on the following factors.

1. Database structure.
   The idea underlying the approach from Part I is that there exists background knowledge (background objects) based on which it is possible to define descriptions of the objects to be analyzed (target objects). Therefore, this framework is dedicated to a typical multi-table database, where essential information on target objects is mainly hidden in additional tables.
   The approach described in Part II provides a relational extension of the standard granular computing framework. It means that the extended framework is more universal in terms of the database structure and can be used for a database consisting of one as well as many tables.
2. Pattern representation.
   Patterns in the first approach can easily be transformed to a typical relational form. Therefore, the framework is preferred to obtain patterns compatible with the original database in terms of the language they are expressed.
   Patterns in the second approach extend propositional ones. This solution makes it easy to upgrade a standard data mining algorithm to a relational case so that the language for expressing patterns is as much as possible close to the original one, i.e. attribute-value language.

The contribution of settlement of relational data mining in the paradigm of granular computing can be summarized into the following main points.

© Springer International Publishing AG 2017   115
P. Hońko, *Granular-Relational Data Mining*, Studies in Computational
Intelligence 702, DOI 10.1007/978-3-319-52751-2_10

1. Application of granular computing approach enables to unify the process of knowledge discovery. Relational information granules are used to build an alternative representation of relational data. This representation also plays the role of the platform for discovering patterns of different types.
2. Richer knowledge can be discovered from relational data transformed into a granular representation. Relation-based granules, which are more informative than granules based on which they are constructed, enable to express knowledge that requires to use additional computations such as aggregation when mining relational data directly.
3. Granular computing based framework can deal with uncertainty in data. The adaptation of one of the main granular computing tools, i.e. rough set theory makes it possible to construct approximate descriptions of concepts defined in singular as well as multi-universes.

Furthermore, the introduced frameworks fill the gap between two research areas: relational data mining and granular computing. They can be considered as the first ones that comprehensively define relational data, information, and knowledge in the paradigm of granular computing.

References

1. Agrawal R, Srikant R (1994) Fast algorithms for mining association rules. In: Bocca JB, Jarke M, Zaniolo C (eds) Proceedings of 20th international conference on very large data bases (VLDB '94). Morgan Kaufmann, San Francisco, pp 487–499
2. Agrawal R, Imieliński T, Swami A (1993) Mining association rules between sets of items in large databases. In: Proceedings of the 1993 ACM SIGMOD international conference on management of data. ACM, New York, NY, USA, SIGMOD '93, pp 207–216
3. Akoglu L, Tong H, Koutra D (2014) Graph-based anomaly detection and description: a survey. CoRR
4. Anderson G, Pfahringer B (2007) Clustering relational data based on randomized propositionalization. In: Inductive logic programming, 17th international conference, ILP 2007, Corvallis, OR, USA, June 19-21, pp 39–48
5. Antonelli M, Ducange P, Lazzerini B, Marcelloni F (2016) Multi-objective evolutionary design of granular rule-based classifiers. Granul Comput 1(1):37–58
6. Apolloni B, Bassis S, Rota J, Galliani GL, Gioia M, Ferrari L (2016) A neurofuzzy algorithm for learning from complex granules. Granul Comput 1–22
7. Appice A, Ceci M, Malgieri C, Malerba D (2007) Discovering relational emerging patterns. In: Basili R, Pazienza MT (eds) AI*IA 2007: artificial intelligence and human-oriented computing: 10th congress of the italian association for artificial intelligence, Rome, Italy, 10-13 Sept 2007. Proceedings, Springer, Berlin, pp 206–217
8. Banks D, House L, McMorris FR, Arabie P, Gaul W (2004) Classification, clustering, and data mining applications. Springer, New York Inc, Secaucus, NJ, USA
9. Bargiela A, Pedrycz W (2003) Granular computing: an introduction. Kluwer Academic Publishers, Boston
10. Bargiela A, Pedrycz W (2008) Toward a theory of granular computing for human-centered information processing. IEEE Trans Fuzzy Syst 16(2):320–330
11. Batagelj V, Ferligoj A (2000) Clustering relational data. In: Gaul W, Opitz O, Schader M (eds) Data analysis: scientific modeling and practical application. Springer, Berlin, pp 3–15
12. Blockeel H, De Raedt L (1997) Relational knowledge discovery in databases. In: Muggleton S (ed) Inductive logic programming: 6th international workshop, ILP-96 Stockholm, Sweden, 26–28 Aug 1996 Selected Papers. Springer, Berlin, pp 199–211
13. Blockeel H, De Raedt L (1998) Top-down induction of first order logical decision trees. Artif Intell 101(1&2):285–297
14. Blockeel H, De Raedt L, Ramon J (1998) Top-down induction of clustering trees. In: Proceedings of the Fifteenth international conference on machine learning. Morgan Kaufmann Publishers Inc., San Francisco, CA, USA, ICML '98, pp 55–63

© Springer International Publishing AG 2017

P. Hońko, *Granular-Relational Data Mining*, Studies in Computational Intelligence 702, DOI 10.1007/978-3-319-52751-2

15. Bonikowski Z, Bryniarski E, Wybraniec-Skardowska U (1998) Extensions and intentions in the rough set theory. Inf Sci 107:149–167
16. Ciucci D (2016) Orthopairs and granular computing. Granul Comput 1(3):159–170
17. De Amo S, Furtado DA (2007) First-order temporal pattern mining with regular expression constraints. Data Knowl Eng 62(3):401–420
18. De Raedt L (2008) Logical and relational learning. Springer, Berlin
19. De Raedt L, Blockeel H, Dehaspe L, van Laer W (2001) Three companions for data mining in first order logic. In: [25]. Springer, pp 105–139
20. Dehaspe L, De Raedt L (1997) Mining association rules in multiple relations. Proceedings of the 7th international workshop on inductive logic programming. Springer, Berlin, pp 125–132
21. Dehaspe L, Toivonen H (1999) Discovery of frequent DATALOG patterns. Data Min Knowl Discov 3(1):7–36
22. Dehaspe L, Toivonen H (2001) Discovery of relational association rules. In: [25]. Springer, pp 189–208
23. Dubois D, Prade H (2016) Bridging gaps between several forms of granular computing. Granul Comput 1(2):115–126
24. Džeroski S, Lavrač N (2001a) An introduction to inductive logic programming. In: [25]. Springer, pp 48–71
25. Džeroski S, Lavrač N (2001b) Relational data mining. Springer, Berlin
26. Džeroski S (2006) From inductive logic programming to relational data mining. In: Fisher M, van der Hoek W, Konev B, Lisitsa A (eds) Proceedings of logics in artificial intelligence: 10th European conference, JELIA 2006 Liverpool, UK, 13–15 Sept 2006. Springer, Berlin, pp 1–14
27. Esposito F, Mauro ND, Basile TMA, Ferilli S (2008) Multi-dimensional relational sequence mining. Fundam Inform 89(1):23–43
28. Ferreira CA, Gama J, Santos Costa V (2012) Predictive sequence miner in ILP learning. In: Muggleton SH, Tamaddoni-Nezhad A, Lisi FA (eds) Inductive logic programming: 21st international conference, ILP 2011, Windsor Great Park, UK, July 31–Aug 3, 2011. Revised Selected Papers, Springer, Berlin, pp 130–144
29. Fonseca NA, Santos Costa V, Camacho R (2012) Conceptual clustering of multi-relational data. In: Muggleton SH, Tamaddoni-Nezhad A, Lisi FA (eds) Inductive logic programming: 21st international conference, ILP 2011, Windsor Great Park, UK, July 31–Aug 3, 2011. Revised Selected Papers, Springer, Berlin, pp 145–159
30. Gazala AH, Ahmad W (2015) Multi-relational data mining a comprehensive survey. In: Usman M (ed) Improving knowledge discovery through the integration of data mining techniques. IGI Global, pp 32–53
31. Gomolińska A (2009) Rough approximation based on weak q-RIFs. Trans Rough Sets 10:117–135
32. Greco S, Matarazzo B, Slowinski R (2001) Rough sets theory for multicriteria decision analysis. Eur J Oper Res 129:1–47
33. Han J, Cheng H, Xin D, Yan X (2007) Frequent pattern mining: current status and future directions. Data Min Knowl Discov 15:55–86
34. Han J, Kamber M, Pei J (2011) Data mining: concepts and techniques, 3rd edn. Morgan Kaufmann Publishers Inc., San Francisco, CA, USA
35. Hipp J, Güntzer U, Nakhaeizadeh G (2000) Algorithms for association rule mining—a general survey and comparison. SIGKDD Explor Newslett 2:58–64
36. Hońko P (2010) Simialrity-based classification in relational databases. Fundam Inform 101(3):187–213
37. Hońko P (2012) Rough-granular computing based relational data mining. In: Greco S, Bouchon-Meunier B, Coletti G, Fedrizzi M, Matarazzo B, Yager RR (eds) Advances on computational intelligence, vol 297. Springer, Berlin, Communications in Computer and Information Science, pp 290–299
38. Hońko P (2013a) Association discovery from relational data via granular computing. Inf Sci 234:136–149

39. Hońko P (2013b) Granular computing for relational data classification. J Intell Inf Syst 41(2):187–210
40. Hońko P (2014) Upgrading a granular computing based data mining framework to a relational case. Int J Intell Syst 29(5):407–438
41. Hońko P (2015a) Description languages for relational information granules. Fundam Inform 137(3):323–340
42. Hońko P (2015b) Relation-based granules to represent relational data and patterns. Appl Soft Comput 37(C):467–478
43. Hońko P (2016a) Compound approximation spaces for relational data. Int J Approx Reason 71:89–111
44. Hońko P (2017) Properties of a granular computing framework for mining relational data. Int J Intell Syst 32(3):227–248
45. Hu X, Pedrycz W, Wang X (2015) Comparative analysis of logic operators: a perspective of statistical testing and granular computing. Int J Approx Reason 66:73–90
46. Karwath A, Kersting K, Landwehr N (2008) Boosting relational sequence alignments. In: Proceedings of the 8th IEEE international conference on data mining (ICDM 2008), 15–19 Dec 2008. Pisa, Italy, pp 857–862
47. Kirsten M, Wrobel S (1998) Relational distance-based clustering. In: Page D (ed) Proceedings of inductive logic programming: 8th international conference, ILP-98 Madison, Wisconsin, USA, 22-24 July 1998. Springer, Berlin, pp 261–270
48. Knobbe AJ (2006) Multi-relational data mining. Fr Art Int 145, IOS Press, Amsterdam, Netherlands
49. Knobbe AJ, Siebes A, Blockeel H, Wallen DVD (2000) Multi-relational data mining, using UML for ILP. In: Principles of data mining and knowledge discovery, pp 1–12
50. Kramer S, Lavrač N, Flach PA (2001) Propositionalization approaches to relational data mining. In: Džeroski S, Lavrač N (eds) Relational data mining. Springer
51. Krogel MA, Rawles S, Zelezny F, Flach P, Lavrač N, Wrobel S (2003) Comparative evaluation of approaches to propositionalization. In: Proceedings of the international conference on inductive logic programming, pp 197–214
52. Lan S, Xiangzhi H (2007) Rough set model with double universe of discourse. In: Proceedings of the IEEE international conference on information reuse and integration, IEEE systems, man, and cybernetics society, pp 492–495
53. Lavrač N, Vavpetič A (2015) Relational and semantic data mining. In: Calimeri F, Ianni G, Truszczynski M (eds) Proceedings of logic programming and nonmonotonic reasoning: 13th international conference, LPNMR 2015, Lexington, KY, USA, 27-30 Sept 2015. Springer International Publishing, Cham, pp 20–31
54. Lavrač N, Džeroski S, Grobelnik M (1991) Learning nonrecursive definitions of relations with LINUS. In: Kodratoff Y (ed) Proceedings of machine learning—EWSL-91: European working session on learning porto, Portugal, 6–8 Mar 1991. Springer, Berlin, pp 265–281
55. Lavrač N, Flach P, Todorovski L (2002) Rule induction for subgroup discovery with cn2-sd. In: Bohanec M, Kasek B, Lavrač N, Mladenic D (eds) ECML/PKDD'02 workshop on integration and collaboration aspects of data mining. University of Helsinki, Decision Support and Meta-Learning, pp 77–87
56. Lavrač N, Železný F, Flach PA (2003) Rsd: Relational subgroup discovery through first-order feature construction. In: Matwin S, Sammut C (eds) Inductive Logic Programming: 12th international conference, ILP 2002 Sydney, Australia, 9–11 July 2002 Revised Papers. Springer, Berlin, pp 149–165
57. Li J, Mei C, Xu W, Qian Y (2015) Concept learning via granular computing: a cognitive viewpoint. Inf Sci 298:447–467
58. Lin TY (2008) Granular computing: common practices and mathematical models. In: Proceedings of IEEE international conference on fuzzy systems (FUZZ-IEEE 2008), IEEE Computer Society, pp 2405–2411
59. Lisi FA, Malerba D (2004) Inducing multi-level association rules from multiple relations. Mach Learn 55:175–210

60. Liu C, Zhong N (2001) Rough problem settings for ILP dealing with imperfect data. Comput Intell 17(3):446–459
61. Livi L, Sadeghian A (2016) Granular computing, computational intelligence, and the analysis of non-geometric input spaces. Granul Comput 1(1):13–20
62. Ma L (2015) Some twin approximation operators on covering approximation spaces. Int J Approx Reason 56:59–70
63. Maciel L, Ballini R, Gomide F (2016) Evolving granular analytics for interval time series forecasting. Granul Comput pp 1–12
64. Maervoet J, Vens C, Berghe GV, Blockeel H, De Causmaecker P (2012) Outlier detection in relational data: a case study in geographical information systems. Expert Syst Appl 39(5):4718–4728
65. Martienne E, Quafafou M (1998) Learning logical descriptions for document understanding: a rough sets-based approach. In: Polkowski L, Skowron A (eds) Rough sets and current trends in computing. Springer, LNCS, pp 202–209
66. Midelfart H, Komorowski HJ (2000) A rough set approach to inductive logic programming. In: Ziarko W, Yao YY (eds) Rough sets and current trends in computing, Springer, LNCS, vol 2005, pp 190–198
67. Milton RS, Maheswari VU, Siromoney A (2004) Rough sets and relational learning. In: Transactions on rough sets I, LNCS, vol 3100. Springer, pp 321–337
68. Milton RS, Maheswari VU, Siromoney A (2005) Studies on rough sets in multiple tables. In: Slezak D, Wang G, Szczuka MS, Duentsch I, Yao Y (eds) RSFDGrC (1). Lecture Notes in Computer Science, vol 3641. Springer, pp 265–274
69. Muggleton S (1991) Inductive logic programming. New Gener Comput 8(4):295–318
70. Muggleton S (1995) Inverse entailment and Progol. New Gener Comput 13(3&4):245–286
71. Pawlak Z (1991) Rough sets., Theoretical aspects of reasoning about dataKluwer Academic, Dordrecht
72. Pedrycz W, Skowron A, Kreinovich V (2008) Handbook of granular computing. Wiley, New York
73. Peters G, Weber R (2016) DCC: a framework for dynamic granular clustering. Granular Comput 1(1):1–11
74. Peters G, Lingras P, Slezak D, Yao Y (eds) (2012) Rough sets: selected methods and applications in management and engineering. Advanced information and knowledge processing. Springer
75. Plotkin GD (1970) A note on inductive generalization. Mach Intell 5:153–163
76. Qian Y, Liang J, Yao Y, Dang C (2010) MGRS: a multi-granulation rough set. Inf Sci 180(6):949–970
77. Quinlan JR, Cameron-Jones RM (1993) FOIL: a midterm report. In: P B (ed) Proceedings of the European conference on machine learning. Springer, pp 3–20
78. Riahi F, Schulte O (2016) Propositionalization for unsupervised outlier detection in multi-relational data. In: Markov Z, Russell I (eds) FLAIRS conference. AAAI Press, pp 448–453
79. Riahi F, Schulte O, Li Q (2014) A proposal for statistical outlier detection in relational structures. In: AAAI workshop: statistical relational artificial intelligence, AAAI, AAAI workshops, vol WS-14-13
80. She Y, He X (2012) On the structure of the multigranulation rough set model. Knowl Based Syst 36:81–92
81. Shen Q, Jensen R (2007) Rough sets, their extensions and applications. Int J Autom Comput 4:217–228
82. Skowron A, Stepaniuk J (1996) Tolerance approximation spaces. Fundam Inform 245–253
83. Skowron A, Stepaniuk J (2001) Information granules: towards foundations of granular computing. Int J Intell Syst 16(1):57–85
84. Skowron A, Stepaniuk J (2004) Constrained sums of information systems. In: Tsumoto S, Slowinski R, Komorowski HJ, Grzymala-Busse JW (eds) Rough sets and current trends in computing, Lecture Notes in Computer Science, vol 3066. Springer, pp 300–309

85. Skowron A, Stepaniuk J, Swiniarski R (2012) Modeling rough granular computing based on approximation spaces. Inf Sci 184(1):20–43
86. Skowron A, Jankowski A, Dutta S (2016) Interactive granular computing. Granular Comput 1(2):95–113
87. Slimani T (2013) Application of rough set theory in data mining. Int J Comput Sci Netw Solutions 1(3):1–10
88. Stepaniuk J (2000) Knowledge discovery by application of rough set models. In: Polkowski ST, Lin T (eds) Rough set methods and applications: new developments in knowledge discovery in information systems. Physica-Verlag, Heidelberg, pp 137–233
89. Stepaniuk J (2008) Rough-granular computing in knowledge discovery and data mining. Stud Comp Intell 152. Springer, Berlin
90. Stepaniuk J, Hońko P (2004) Learning first-order rules: a rough set approach. Fundam Inform 61:139–157
91. Tan A, Li J, Lin Y, Lin G (2015) Matrix-based set approximations and reductions in covering decision information systems. Int J Approximate Reasoning 59:68–80
92. Tan PN, Steinbach M, Kumar V (2005) Introduction to data mining, 1st edn. Addison-Wesley Longman Publishing Co. Inc, Boston
93. Thangaraj DM, Vijayalakshmi C (2011) A study on classification approaches across multiple database relations. Int J Comput Appl 12(12):1–6
94. Vaghela VB, Vandra KH, Modi NK (2012) Analysis and comparative study of classifiers for relational data mining. Int J Comput Appl 55:11–21
95. Van Laer W, De Raedt L (2001) How to upgrade propositional learners to first order logic: a case study. In: [25]. Springer, pp 235–261
96. Wang L, Liu X, Qiu W (2012) Nearness approximation space based on axiomatic fuzzy sets. Int J Approximate Reasoning 53(2):200–211
97. Wilke G, Portmann E (2016a) Granular computing as a basis of human-data interaction: a cognitive cities use case. Granular Comput 1(3):181–197
98. Wilke G, Portmann E (2016b) Granular computing as a basis of human–data interaction: a cognitive cities use case. Granular Comput 1–17
99. Wrobel S (1997) An algorithm for multi-relational discovery of subgroups. In: Komorowski J, Zytkow J (eds) Principles of data mining and knowledge discovery: first European symposium, PKDD '97 Trondheim, Norway, June 24–27, 1997 proceedings. Springer, Berlin, pp 78–87
100. Xu Z, Wang H (2016) Managing multi-granularity linguistic information in qualitative group decision making: an overview. Granular Comput 1(1):21–35
101. Yan R, Zheng J, Liu J, Zhai Y (2010) Research on the model of rough set over dual-universes. Knowl Based Syst 23(8):817–822
102. Yang Q, Wu X (2006) 10 challenging problems in data mining research. Int J Inf Tech Decis 5(4):597–604
103. Yao JT (2005) Information granulation and granular relationships. In: Hu X, Liu Q, Skowron A, Lin TY, Yager RR, Zhang B (eds) Proceedings of the IEEE conference on granular computing. IEEE Computer Society, pp 326–329
104. Yao JT, Vasilakos AV, Pedrycz W (2013) Granular computing: perspectives and challenges. IEEE T Cybern 43(6):1977–1989
105. Yao YY (2000) Granular computing: basic issues and possible solutions. In: Wang P (ed) Proceedings of the 5th joint conference on information sciences (JCIS). Association for Intelligent Machinery, pp 186–189
106. Yao YY (2004) A comparative study of formal concept analysis and rough set theory in data analysis. In: Tsumoto S, Slowinski R, Komorowski HJ, Grzymala-Busse JW (eds) Rough sets and current trends in computing, Lecture Notes in Computer Science, vol 3066. Springer, pp 59–68
107. Yao YY (2007) Decision-theoretic rough set models. In: Rough sets and knowledge technology, second international conference, RSKT 2007, Toronto, Canada, May 14–16, 2007, proceedings, pp 1–12

108. Yin X, Han J, Yang J (2003) Efficient multi-relational classification by tuple id propagation. In: Džeroski S, De Raedt L, Wrobel S (eds) Proceedings of the second international workshop on multi-relational data mining (MRDM-2003). ACM Press, pp 122–134

109. Zadeh LA (1965) Fuzzy sets. Inf Control 8:338–353

110. Zadeh LA (1997) Towards a theory of fuzzy information granulation and its centrality in human reasoning and fuzzy logic. Fuzzy Set Syst 90(2):111–127

111. Železný F, Lavrač N (2006) Propositionalization-based relational subgroup discovery with RSD. Mach Learn 62(1):33–63

112. Zhang J, Li T, Chen H (2012) Composite rough sets. In: Lei J, Wang FL, Deng H, Miao D (eds) AICI, Lecture Notes in computer science, vol 7530. Springer, pp 150–159

113. Zhang J, Li T, Chen H (2014) Composite rough sets for dynamic data mining. Inf Sci 257:81–100

114. Zhen P, Wu L, Wang X (2009) Research on multi-relational classification approaches. In: Proceedings of the 2009 international conference on computational intelligence and natural computing, vol 01. IEEE Computer Society, Washington, pp 51–54

115. Zhu W, Wang F (2003) Reduction and axiomatization of covering generalized rough sets. Inf Sci 152:217–230

116. Ziarko W (1993) Variable precision rough set model. J Comput Syst Sci 46(1):39–59

Index

Printed in the United States
By Bookmasters